A. Ritchie Leask

Triple and Quadruple Expansion Engines

And Boilers and Their Management

A. Ritchie Leask

Triple and Quadruple Expansion Engines
And Boilers and Their Management

ISBN/EAN: 9783337139209

Printed in Europe, USA, Canada, Australia, Japan

Cover: Foto ©berggeist007 / pixelio.de

More available books at **www.hansebooks.com**

TRIPLE AND QUADRUPLE
EXPANSION ENGINES AND BOILERS

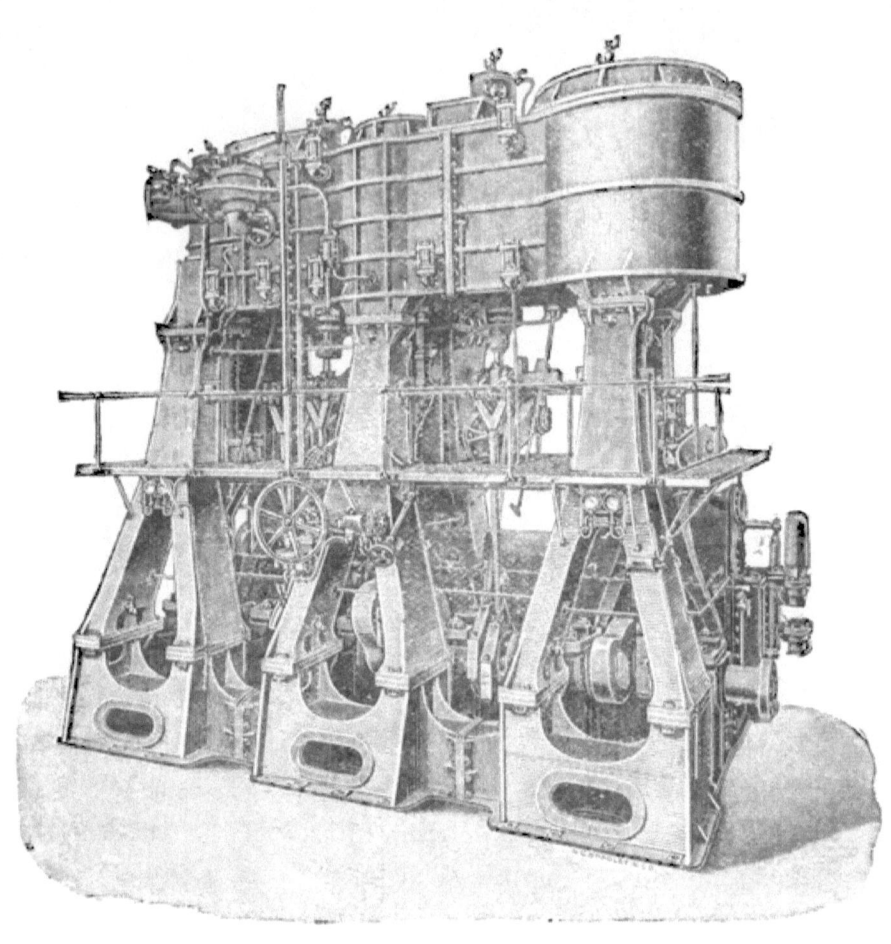

TRIPLE EXPANSION ENGINES OF P. & O. S.S. "ADEN."

TRIPLE & QUADRUPLE EXPANSION ENGINES & BOILERS

AND THEIR MANAGEMENT

WITH FIFTY-NINE ILLUSTRATIONS

By A. RITCHIE LEASK

AUTHOR OF
"REFRIGERATING MACHINERY: ITS PRINCIPLES AND MANAGEMENT"

TOWER PUBLISHING CO., 91 MINORIES, LONDON, E.
W. & T. ROUTLEY, 22 & 24 ERSKINE ST., SYDNEY, N.S.W.
1892

MORRISON AND GIBB, PRINTERS, EDINBURGH.

PREFACE.

THIS work has been written to meet a widely-expressed desire for information regarding the management of triple and quadruple expansion engines and boilers, and the Author has endeavoured to produce a book that would, in plain everyday language, place before engineers a digest of the experience of those who have been in charge of machinery of the above description, together with such rules and directions as may have been suggested by these experiences, in addition to a popular description of the more prominent inventions and appliances that have been recently adopted in connection with such machinery.

For much of the technical information contained in this work, the Author is indebted to the papers on "Forced Draught," read by Mr. James Howden before the Institute of Naval Architects in 1884 and 1886; to a similar paper read by Mr. A. E. Seaton before the Hull and District Institute of Engineers and Shipbuilders in

1891; to Mr. B. H. Thwaite, for the greater part of the chapter on "Liquid Fuel;" to Professor Vivian B. Lewes, for most of the chapter on "Pitting and Corrosion of Boilers," and which has been taken by permission from his valuable work entitled *Service Chemistry*, and also for his kind assistance in revising the proofs of that chapter. Messrs. T. Richardson & Sons have kindly contributed the specifications of a set of triple expansion engines and boilers, besides the description and illustration of Morison's Patent Evaporator and Feed-Heater, and a photograph of the engines of the P. and O. Mail steamship *Aden*. The original papers that have appeared in the *Engineers' Gazette* from time to time have also been drawn upon, and various other contributions quoted, but in nearly every case the sources have been mentioned.

In recognition of the eminent services rendered by Dr. A. C. KIRK to the engineering profession, particularly by his introduction of the triple expansion engine, and in gratitude to him for his generous assistance in revising the proof-sheets of this work and furnishing many valuable suggestions and items of information, it has been respectfully dedicated to him by

THE AUTHOR.

CONTENTS.

CHAPTER I.

	PAGE
Introduction—The First Triple Expansion Engine—The Advantages of Triple Expansion Engines—Difficulties in the Way of their Adoption—Improvements in Machinery and Appliances for Boiler-Making—Saving Effected by Employing Triple Expansion Engines—Objections to their Use—Quadruple Expansion Engines — Their Advantages and Disadvantages—Summary, ...	1–9

CHAPTER II.

Loss of Power through Friction—Pistons and Packing-Rings—How they should be Fitted—Compression required—Excessive Pressure of some Springs—Devices intended to obviate this—Solid Packing-Rings—Ramsbottom Rings—Coach Springs—Modern Coil Springs—Their Advantages—How to apply them, ... 10–20

CHAPTER III.

The Slide Valve—Importance of keeping it in Order—Arrangements for Reducing Friction — The Piston Valve — Its Advantages and Disadvantages—Haythorn's Patent Piston Valve—Radial Valve-Gears—Reason for their Adoption—Hackworth's, Marshall's, Bremme's, Kitson's, Kirk's, Bryce Douglas', Joy's, Walschaert's, Morton's, Brown's, Sisson's, and Howden's Radial Valve-Gears — Requirements of a Theoretically Perfect Slide Valve-Gear—Comparison of the Different Valve-Gears, ... 21–43

CHAPTER IV.

The Sequence of Cranks—Greater Economy with the High-Pressure Crank Leading—Proportions for Cylinder Diameters—Three Cranks better than Two—The Steam-Jacket—Its Advantages or otherwise—Most Economical on Intermediate and Low-Pressure Cylinders—Friction of Engines—Experiments thereon—How to Reduce it—Packing of Glands—Metallic Packings—Duval's—United States—Macbeth's Patent—The Ideal Packing, ... 44–60

CHAPTER V.

Lining and Adjustment of Bearings—Thrust Collars—Air-Pump—Circulating-Pump—Feed-Pumps—Condenser—Lubrication of Working Parts—Caution as to injurious Oils—Sight-feed Lubricator—Expansion Links—How to alter the HP by their Means—Feed-water Heaters—Explanation of their Economy—A Theoretically Perfect Feed-Heater—Weir's Feed-Heater and Evaporator—Morison's Evaporator and Feed-Heater—Board of Trade and Lloyds' Requirements, . 61–107

CHAPTER VI.

Combustion of Fuel—Economy of Cornish Boilers—How Combustion is Effected—Quantity of Air Required—Natural Draught—Forced Draught—Induced Draught—Different Methods of Effecting the Same—Their Respective Advantages and Disadvantages—Howden's System of Forced Draught—Results obtained from Actual Practice, . . . 108–150

CHAPTER VII.

Liquid Fuel—Its Early Introductions—Geographical Distribution—Theories as to its Origin—Chemical Constitution—Thermic Value compared with that of Coal—Possible Efficiency—Cost of Oil to admit of its Economical Use—The Value of Increased Cargo Space—Dangers of Stowage—Precautions to be observed in Oil Storage—Methods of Application—Injectors—Use of Oil under Pressure—Forced Draught Oil Firing System—Tabulated Result of Trials—Advantages and Disadvantages—Henwood's System of Oil Firing, ... 151–164

Contents. xiii

CHAPTER VIII.

The Marine Boiler—Quantity of Steam Generated and Fuel Consumed—Perfect Combustion—Size of Fire-Bars—Length of Fire-Grate—Height of Bridges—Circulation in a Boiler—Prevention of Scale—Constituents of Scale—Evil Effects of Scale—Collapsed Furnaces—Edmiston's Marine Filter—Corrosion and Pitting—How Caused—Chemical Composition of Corroded Parts—Methods of preventing Corrosion—Hannay's Patent Electrogen, ... 165–214

CHAPTER IX.

Management of Working Surfaces—Crank Shaft, Crank Pin, Pump Lever Brasses—Liners—Stripping Brasses—Bearing Surface—Hints on Adjusting Brasses—Care of Built Crank Shafts—Guide Bars and Shoes—How to Lubricate them—The Thrust Block—Causes of Trouble and their Remedies—Adjustment of Tunnel Bearings—Warming up Engines—Making Ready to Start—Precautions under Way—Sweeping Tubes—Cleaning Fires—Blowing down Boilers—Treatment of Ashes—Fuel Economy, ... 215–234

CHAPTER X.

Specification of Triple Expansion Marine Engines—Cylinders—Cylinder Covers—Pistons—Slide Valves—Valve-Gear—Piston-Rods—Connecting-Rods—Columns—Air-Pump—Circulating-Pump—Feed-Pumps—Bilge-Pumps—Condenser—Evaporator—Drain Trap—Bed-plate—Crank Shaft—Shafting—Propeller—Turning Gear—Telegraph—Starting Gear—Throttle Valve—Donkey Engine—Sanitary Pump—Ballast Engine—Pipes, Lubricators, etc.—Conditions—Boilers—Tubes—Staying—Funnel and Uptake—Mountings—Ventilators, ... 235–246

LIST OF ILLUSTRATIONS.

FIGS		PAGE
	Frontispiece,	vi
1.	Adjustable Packing-Ring,	12
2.	Haythorn's Patent Piston,	15
3.	The Lancaster Spiral Coil,	17
4.	The Lancaster Serpent Coil,	18
5.	Lancaster & Tonge's Patent Piston-Block,	18
6.	Rickaby's Patent Piston-Spring,	19
7.	Church's Equilibrium Ring, Section,	22
8.	,, ,, Plan,	23
9.	Haythorn's Patent Piston Valve,	26
10–12.	Hackworth's Radial Valve-Gear,	28
13, 14.	Marshall's ,,	30
15.	Bremme's ,,	31
16.	Kitson's ,,	31
17.	Kirk's ,,	32
18.	Bryce Douglas' ,,	32
19.	Joy's ,,	34
20.	Walschaert's ,,	35
21.	Morton's, ,,	36
22.	Brown's ,,	37
23.	Sisson's ,,	39
24, 25.	Howden's ,,	40, 42

		PAGE
26-28. United States Metallic Packing,	. . . .	53, 55
29. Macbeth's Patent ,,	.	56
30. Nuts with Divisions for Adjusting,	. . .	62
31-33. Weir's Feed-Heater,	. . .	78-80
34, 35. Rayner's Automatic Evaporator,	. .	83, 86
36. Morison's Patent Evaporator,	. . .	93
37, 38. Sectional Elevation and Plan of same,	.	95, 96
39. Morison's Feed-Water Heater,	. .	98
40. ,, Reducing Valve,	.	100
41, 42. ,, Brining Cock,	. .	101
43. Evaporating into Condenser,	. . .	102
44. ,, I.P. Valve Casing and Condenser,		104
45. ,, Feed-Heater,	. . .	105
46. Howden's System of Forced Draught,	.	128
47. Thwaite's System of Liquid Fuel,	. .	160
48. Edmiston's Marine Filter,	. .	182
49. Hannay's Electrogen,	. .	207
50, 51. Brasses cut away at the Sides,	.	216
52. Gudgeons cut away at the Sides,	.	217
53. Approved Method of cutting Oil Ways,	.	217
54. Another ,, ,,	.	218
55. Main Brasses eased at the Sides,	.	219
56. Illustration of badly-adjusted Brasses,	.	220
57. Imperfect Method of Guide Lubrication,		221
58. Approved Method of Shoe Lubrication,	. .	221
59. Ashes Washing Machine, .	.	232

TRIPLE AND QUADRUPLE EXPANSION ENGINES AND BOILERS AND THEIR MANAGEMENT.

CHAPTER I.

Introduction—The First Triple Expansion Engine—The Advantages of Triple Expansion Engines—Difficulties in the Way of their Adoption—Improvements in Machinery and Appliances for Boiler-Making—Saving Effected by Employing Triple Expansion Engines—Objections to their Use—Quadruple Expansion Engines—Their Advantages and Disadvantages—Summary.

THE rapid development of the marine engine of late has not only placed additional burdens and responsibilities upon engineers, but has rendered it imperative that they should acquire a knowledge of triple and quadruple expansion of steam, and in many cases also of forced draught and the use of liquid fuel. It therefore becomes the duty of every engineer, who desires to keep in the front rank of his profession, to increase his scientific knowledge to such an extent as will enable him to deal intelligently with the improved applications of energy and the altered conditions now presented to him.

No doubt there are numerous text-books on the steam engine from which an engineer may obtain a deal of scientific information, but it is quite as essential that he should possess *practical* as well as scientific knowledge regarding the latest developments of the marine engine. There being no book extant which treats of triple and quadruple expansion engines, from a practical point of view at least, it is hoped that the following particulars regarding the latest types of engines, valve-gears, and boilers, and the most approved method of working them, written in a plain and unpretentious style, and embodying the recent experiences of engineers who have sailed with such engines, may be found of service to our sea-going brethren, and meet with their approval.

Being written so that it may be easily understood by all our readers, it is believed that engine-fitters and apprentices who intend going to sea would also find it to be to their advantage to study the record of experiences set forth in this treatise, as the knowledge thereby acquired would materially assist them in discharging their future duties on board ship, and render their services much more acceptable to their chief and fellow engineers than they would otherwise be.

It is only within the last few years that triple expansion engines may be said to have come into general use, but as early as 1874 Mr. W. H. Dixon, a Liverpool shipowner, being fully alive to the advantages that had been gained by the compound over the common engine with increased steam pressure, resolved to go a step further, and he accordingly instructed Messrs. John Elder & Co., the well-known engineers and shipbuilders, to replace the machinery of his s.s. *Propontis* by new machinery, the boilers to be water-tube boilers on Rowan & Horton's patent, for a working pressure of 150 lbs. per square inch. In these boilers the water was contained *inside* a number of wrought-iron tubes, which were of small diameter and, consequently, of great strength, while the fire acted on the outside.

Mr. A. C. Kirk was entrusted with the designing of the necessary machinery for utilising steam of this high pressure, and in order to obtain the best results he found that it would be necessary to expand the steam in three cylinders successively instead of two as in the compound engine. By this arrangement the variation of temperature in each cylinder would only be two-thirds of that in a two-cylinder engine, while the advantages to be gained by it may be summed up as follows: (a) a higher rate of expansion; (b) the loss from variation of temperature in the cylinders would be less; (c) the initial strains would be less, which would admit of a corresponding decrease in the size of the working parts.

The engines were of the ordinary inverted type, with three cranks, and the cylinders were 23 inches, $41\frac{1}{4}$ inches, and $62\frac{1}{8}$ inches diameter respectively; and it may be interesting to learn that they gave every satisfaction. It was unfortunately different with the boilers, however, as the chambers above the fires of two of them were burnt and afterwards burst. They were then taken out and the ordinary type of boilers with a working pressure of 90 lbs. substituted, with which the engines worked satisfactorily, but not, of course, with the same economy as with the higher pressure. After the use of high-pressure steam in boilers of the ordinary type was established, these boilers were removed and replaced by boilers suited to the original pressure of 150 lbs. Owing to the failure of the water-tube boilers, shipowners were unwilling to run the risk of using steam at a high pressure, and it was not until 1877 that a further attempt was made. It was then principally owing to the great improvements in the manufacture of mild steel that boilers of the ordinary type were constructed of this material, its superior strength enabling boiler-makers to increase the pressure 30 per cent. for the same thickness of plate, or, what is equivalent to this, for an equal pressure the thickness of plate could be reduced 25 per cent.

About this time also great improvements were made in the machinery and appliances used in boiler-making, notably those for the bending, flanging, and rivetting of plates, whereby thicker plates could be worked than had previously been possible, and the adoption of methods of stiffening the furnaces, notably the introduction of Fox's corrugated furnace, in addition to the other improvements mentioned above, admitted of the construction of boilers to withstand a much higher pressure than before.

The saving effected by the adoption of the triple expansion type of engine, so far as can be learned from the performances of a comparatively limited number of steamers, varies from 20 to 30 per cent., consequently 25 per cent. may be taken as the average saving. This is a matter of the utmost importance to the shipowner, and it is not surprising therefore to learn that nearly every steamer now in course of construction, more especially those intended for long voyages, is being fitted with this type of engine.

It was at first considered by some that the durability of such boilers would be much inferior to that exhibited by those working at a lower pressure, and would, moreover, entail a greatly increased expenditure in order to maintain them in a state of efficiency. Judging from the condition of the boilers of a number of steamers that have been running for some years, and the comparatively small amount of repairs they have required during that time, these objections do not appear to be well founded. On the other hand, taking into consideration the fact that the improvements in the machinery and other appliances, and also in the construction of marine boilers, have fully kept pace with the increase of pressure, it is not too much to expect that the boilers now constructed will last nearly, if not quite, as long as those made in previous years.

The greater space occupied by the triple expansion engine is another objection often urged against its adoption, it being evident that three cranks must take up more room in a fore and aft direction than two. As a set-off to this, it must be remembered that the consumption of fuel is from 20 to 30 per cent. less than in the compound engine, which admits of a corresponding reduction in the size of, and consequently in the space occupied by, the boilers. Again, since the consumption of fuel is from 20 to 30 per cent. smaller, the bunkers may be reduced in size to a similar extent, and in many cases, where there are athwartship bunkers, they may be dispensed with, and the space occupied by them utilised in lengthening the engine-room. Some builders also construct triple expansion engines with the valves at the back, indeed those of the *Propontis* were arranged so, and this arrangement is much used at present, while in some cases they are placed in front of instead of between the cylinders, thus bringing them closer together; by this arrangement the three cranks only occupy about the same space as that previously required for two.

This plan is usually adopted in steamers where the old engines are to be replaced by triple expansion, as it entails the minimum amount of alteration in the engine-room, and in most cases renders unnecessary the shifting of bulkheads and bunkers. From the foregoing it will be readily seen that the space occupied by the machinery and bunkers exceeds very little, if any, that previously occupied by the two crank compound engines.

The success of the triple expansion engine has induced some engineers to build engines of the quadruple expansion type, employing an average boiler pressure of 180 lbs. per square inch. A number of leading engineers are of opinion, however, that these engines will not come into as general use as the former, the reasons assigned being that the advantages to be gained are not considered sufficient to compensate for the additional expense, unless the pressure is at least 200 lbs. But we have as yet little or no experience of how the present materials of which boilers or cylinder are made will behave under this high temperature and pressure. The tendency is towards the use of higher pressure, and gradual experience is the safest guide. As to their advantage, the additional power gained by using steam above 150 lbs. pressure is theoretically very small, and it is extremely probable that any gain would be more than counter-balanced by losses from leakage, etc., and four cylinders being employed instead of three there would be a further loss from increased friction. In one case that has come under notice of quadruple expansion engines working at a pressure of 190 lbs. per square inch, it is asserted that the consumption has been as low as 1·25 lbs. per indicated horse-power per hour, but it would be necessary to have reliable data of the vessel's performance during several ordinary voyages before this exceedingly small consumption could be accepted as trustworthy. Another difficulty would be to construct boilers of the present design to withstand much higher pressure than those now in vogue. The quadruple expansion type has its uses, however, as the ordinary compound engine with two cranks may be easily converted into one, in which the steam expands in four cylinders successively, by placing two new cylinders on top of the two old ones, the high pressure being above the third and the second above the low pressure cylinder, thus making of it a two crank tandem engine.

Notwithstanding what has been said regarding the relatively small gain in economy to be derived from the additional cylinder, and the questionable advantages of high steam pressure, the adoption of quadruple engines appears to be slowly yet surely increasing, and instances are not wanting which go to show that quadruple expansion may prove more advantageous than is frequently assumed to be possible. An interesting comparison between triple and quadruple expansion engines has lately been made by Messrs. Yarrow, of Poplar, who have constructed six first-class torpedo boats for the Argentine Government. Five of these vessels were fitted with triple engines and the sixth with quadruple expansion engines. These engines have four inverted cylinders acting on four cranks on one shaft. The cylinders are respectively 14 inches, 20 inches, 27 inches, and 36 inches in diameter, by 16 inches stroke. The engines are placed in pairs, each pair having its cranks opposite one another, which arrangement balances the moving parts and reduces the vibration of the vessel. The boilers of the whole set of vessels are of the locomotive type, having copper fireboxes and brass tubes. At the official trial six runs on the measured mile were made, and a continuous run of two hours. The result gave an average speed of 24·426 knots on the measured mile runs, and an average speed of 24·4 knots on the two hours' continuous run. The average number of revolutions on the measured mile runs was 433 per minute, and on the two hours' run 432 per minute; the average steam pressure being 199 lbs. The average indicated horse-power of the five first-class boats with triple expansion engines was found to be 1120, while the quadruple expansion engines developed 1230 indicated horse-power, showing a gain of 110 horse-power, or 10 per cent., the consumption of fuel being the same.

Quadrupling also affords a very convenient means of increasing the power of compound engines. A practical example of this is supplied by Mr. R. Carson, of Hull, who mentions a vessel, the engines of which were of an economical compound type, indicating about 480 horse-power, the average consumption of coal for the voyage from Hull to Amsterdam and home being 32 tons. After quadrupling, and replacing the boiler by one of his own special design, the engines indicated 540 horse-power, and the consumption of fuel was reduced to 20 tons per voyage for all purposes, thus making a saving of 12 tons per voyage. He accordingly advises the shipowner, who is despondent over being the possessor of a steamer, the engines of which are becoming obsolete, to quadruple his old engines, and thereby obtain the necessary economy to enable the old vessels to compete with those of more modern type. Of course, in this instance, much of the improvement—probably the major portion — would be due to the boiler, but evidently the gain due to quadruple expansion is such that under certain circumstances it may be adopted with advantage.

Having briefly described the most recent types of engines in use at the present time, it may be interesting to sum up the advantages obtained by using steam of from 140 lbs. to 160 lbs. per square inch. The average consumption of good compound engines is about 2 lbs. per indicated horse-power per hour, while with triple expansion engines the average is about $1\frac{1}{2}$ lbs. per hour, or a saving of 25 per cent. By employing three cranks the engines run much more smoothly, thus admitting of a higher speed of piston, and consequently a smaller area of cylinder is required inversely proportioned to the increase of speed. Again, by adopting three cranks the initial strains are lessened, thereby enabling smaller working parts to be substituted, so that triple and compound engines weigh practically the same per horse-power; but, the consumption of fuel being less for a given power and speed, the vessel is able to carry proportionately more cargo.

CHAPTER II.

Loss of Power through Friction—Pistons and Packing-Rings—How they should be Fitted—Compression required—Excessive Pressure of some Springs—Devices intended to obviate this—Solid Packing-Rings—Ramsbottom Rings—Coach Springs—Modern Coil Springs—Their Advantages—How to apply them.

It is an indisputable fact that every portion of machinery in motion necessarily absorbs a certain quantity of power. Every moving joint abstracts its quota, which might otherwise be doing useful work in propelling a ship; and engineers have striven with more or less success to reduce this loss to a minimum, by simplifying parts, and by making improvements in the details of construction.

There are two parts of the steam engine (and these, perhaps, the most important of any) in which comparatively little progress has been made, notwithstanding the many ingenious devices that have appeared from time to time—viz. the piston and steam distributing valves. Of the loss of power through friction of these moving parts some idea may be formed by the amount of wear which takes place in packing-rings and valve-faces. An apparatus was constructed in the United States some time ago for testing slide valves (described in *Engineering*, August 20, 1886), when it was found that from $4\frac{1}{2}$ to $7\frac{1}{2}$ per cent. of the total power shown by the indicator diagram was required to work the slide valve alone, and doubtless thousands of pistons are now working with equally bad results. The cause is not far to seek. The idea that packing-rings should possess unlimited elasticity goes far to account for it. They are consequently often backed up by springs, which would follow up the bore of the cylinder were it even half an inch larger at one end than the other. The result of this excessive pressure and elasticity is enormous friction and rapid destruction of both cylinder and piston-ring, while the gradual formation of hollows in the softer portions of the cylinders intensifies the evil.

Pistons should be made steam-tight, with *the least possible lateral pressure*, a very small amount of compression being quite sufficient to make them steam-tight. The excessive pressure of packing-rings is sometimes due to steam getting into the space behind the packing-ring. In the older forms this was attempted to be obviated by carefully scraping the packing-ring so as to be a floating yet steam-tight fit between the flange of the piston and the junk-ring. Many patent devices have been introduced to secure this, such as fitting two packing-rings forced apart by some arrangement of springs, so as to maintain a steam-tight joint between the packing-ring and the piston and junk-ring. But after working some little time, steam ultimately finds its way behind the ring and presses it against the cylinder, thereby defeating the object aimed at.

In the case of piston valves, one of the most successful fittings is that in which the packing-ring is simply an adjustable solid plug, and this has been in use for some time in large first-class steamers. It may be mentioned that the same thing was done, but not in quite so simple a form, very many years ago by Seaward, of London, and was adopted by the Pacific Steam Navigation Co., the ring being bolted at the joint (Fig. 1), and a distance plate fitted in to make it a swimming fit. This device has also been applied to pistons. Otherwise, packing-rings are made very stiff, so that the steam may not, after a slight wear, press them too hard against the cylinder; and when they are made stiff in this way, they have usually one or more rectangular grooves, from $\frac{1}{4}$ to $\frac{3}{8}$

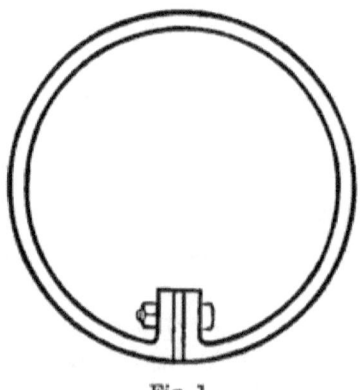

Fig. 1.

inch diameter, carried all the way round their outside surfaces, parallel to the piston-flange, in each of which is fitted a ring of square section made of hammered or spring-tempered brass, and termed a Ramsbottom ring, after its inventor, who first applied this device to locomotive engines. They make an excellent packing, and are much used for high-pressure and intermediate pistons. One important point is that the packing-rings should not bear in any way against the piston, so that while the piston-rod follows its own path, the rings may freely follow the inside of the cylinder.

With the old coach-spring arrangement it was endeavoured to secure this by making the springs abut against a solid floating-ring, concentric with the packing-ring, but these springs were in some cases set to give such an excessive pressure as actually to reduce the revolutions of the engine. The packing-ring should be sprung out to fit the cylinder, and the coach springs "set" in such a manner that they may be readily pushed down into their places by hand; and, in the various forms of piston we now proceed to describe, this point will be found to be attended to.

These springs are now practically discarded in favour of the modern coil springs, having both a lateral and vertical action; and as the packing-rings, if a good fit, require only a very small amount of compression, care should be taken that the junk-ring does not compress them too much, as even with this class of spring an enormous pressure can be put upon the cylinder walls. Although it is most important that the packing-rings should not be too tightly set up, it is equally important that they should not be too slack, for if steam is permitted for any length of time to pass between either the packing and junk rings, or the packing-ring and piston, the exposed surfaces become very rough and irregular, owing to the action of the steam upon them. The consequence will be that, however tightly the packing-ring may afterwards be set up by means of the springs, it will pass steam owing to the roughness and irregularity of the working surfaces. This is perhaps not of so much consequence in the high-pressure cylinder, as the steam that

14   *Triple and Quadruple Expansion Engines*

passes will be utilised in the low-pressure cylinder; but if the low-pressure piston is passing steam it becomes a serious matter, for, in addition to the loss caused by the steam rushing straight into the condenser instead of doing useful work, the live steam will impair the vacuum, and throw more work on the circulating pump, on account of more circulating water being required to condense that steam, and consequently the revolutions of the engine will be reduced.

As many engineers are still unacquainted with the various forms of patent pistons and rings now in use, it may perhaps be advisable to describe one or two of those that are most approved of, and the methods of adjusting them as recommended by their makers. Haythorn's patent piston, manufactured by Messrs. Haythorn & Stuart, of Eastwood Engine Works, Pollokshaws, Glasgow, who are the patentees and sole makers, has been in use for over five years, under all conditions of working, and in various sizes up to nearly 100 inches in diameter, and the patentees guarantee to the users of these pistons an economy, or increased effective power, of at least 5 per cent. in each instance, besides the great saving in repairs, renewals, and disuse of lubricants. Fig. 2 is a plan, section, and elevation of an ordinary marine piston for inverted cylinders. The packing consists of two strong rings of L section; these are expanded radially by a chain consisting of flat wrought-iron links, and double cast-iron or brass pads alternately. At opposite sides these links give place to—on one side a strong spiral spring contained in two half boxes, and on the other an adjusting screw and nut. The link pads are in halves vertically, and between these, threaded on their brass connecting pins, and between one of them and the wrought-iron link is a short spiral spring, separating the half pads against the flanges of packing-rings with a constant pressure. The adjusting screw forces the links

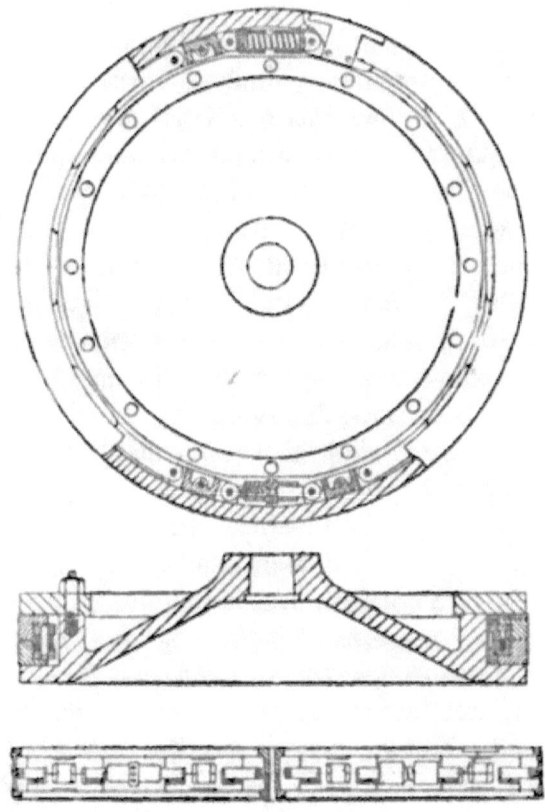

Fig. 2.

round against the spring opposite, compressing it a measurable distance within the half boxes, never more than $\frac{3}{16}$ to $\frac{1}{4}$ inch for the largest diameters. It will thus be seen that the vertical action is constant, and just sufficient to produce steam-tight contact between the packing-rings, junk-ring, and piston-flange, and that the expanding pressure being applied through jointed links is transmitted radially and equally all round the packing-rings.

There is a modification suitable for smaller diameters, and for high-pressure cylinders. In this case the packing-ring is in one piece, having its ends prolonged and with shallow grooves formed on outside to contain a water packing. In this design there are no springs either for radial or vertical pressure. The rings are expanded until they exactly fit the bore of cylinder, and the junk-ring is then screwed down, thus making it equivalent to a solid block. When this is applied to tandem, triple, or quadruple engines, in the upper cylinder it forms the best possible guide for the rod, the lower cylinders in this instance having floating-rings, as in Fig. 2. This arrangement is said to have been particularly successful, and in instances where it has replaced ordinary packing that had been wearing as much as $\frac{1}{2}$ inch open in a voyage its wear has been quite inappreciable.

Another very successful patent is that manufactured by Messrs. Lancaster & Tonge, engineers, Pendleton, near Manchester, and known as the Lancaster patent piston. This is made in two forms, the Lancaster spiral coil piston and the Lancaster patent serpent coil piston. The spiral coil spring, Fig. 3, has two actions: first, the continuous effort of the straight spiral spring when forced into the circular rings to recover its original form; second, the straight spiral spring being diametrically compressed when forced into the rings, the continuous effort to recover its original diameter. The lateral pressure is thus obtained by forcing the rings against the sides of the cylinder, and the vertical pressure by the rings being grooved to the circle of the spring, thus forcing the rings against the block and junk-ring.

The advantages claimed for this spring are that it is simple and reliable even in unskilled hands, as, the vertical pressure being obtained from the lateral, too much friction cannot be exerted against the sides of the cylinder in screwing down the junk-ring. It is self-adjusting and perfectly steam-tight, with a minimum of friction.

Fig. 3.

The method of putting in this spring is as follows: Place the ends of the spring together in the cylinder, holding them firmly whilst forcing in the spring. When taking the spring out, hold the ends firmly, draw the opposite part of the spring to the end of the cylinder, then gradually draw out one end of the spring, keeping tight hold of the other. It must be borne in mind that this spring will always resume its straight form.

The Lancaster serpent coil differs from the spiral coil, as will be seen in the illustration, Fig. 4. It is a round section of tempered steel, and the makers claim that it has many advantages over the universally known style advertised by many engineers, all made from a flat section, so that unless very accurately fitted the piston becomes to all intents and purposes a solid piston, an impossibility with the "serpent," as the round coil

Fig. 4.

against a flat surface cannot bind, but causes a revolving tendency in the packing-rings. There is no difficulty in putting them in place; the rings being slightly rounded,

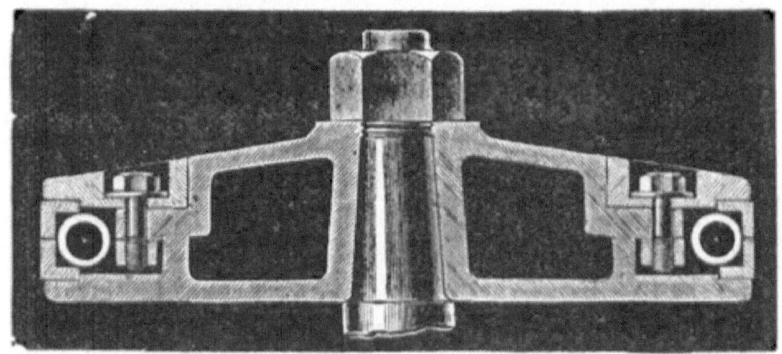

Fig. 5.

the round coil slips in with the greatest ease. Messrs. Lancaster & Tonge have also a patent piston-block, Fig. 5, of which the following is a description: they are fitted with their patent springs for keeping the body of the block from the cylinder and the rings from being driven inwards, thus obviating the uneven wearing of the cylinder by the block bearing on it. This is prevented by turning the grooves shown in the junk-ring and bottom flange so that the rings rest on them, but, being free to expand, keep the piston perfectly steam-tight. One important fact in connection with this patent is that, after boring out a cylinder, the old block can be used by turning down the grooves in the junk-ring and bottom flange to the exact depth required, and the rings and springs, by standing prominent, will be thoroughly efficient, thus saving the expense of a new block.

Another form of piston-spring is that known as Rickaby's patent improved self-adjusting metallic outer ring, made by A. A. Rickaby, Bloomfield Engine Works, Sunderland. It consists essentially of a spiral coil spring of steel, as shown

Fig. 6.

in Fig. 6, which sufficiently explains its construction. The advantages claimed for it are: it can be fitted to any existing piston-block, being made by a patent machine; it is ready for use without turning or chasing; and, being of steel, it is much more durable, elastic, and less liable to corrode by the action of the condensed water and steam. The rings being self-adjusting are consequently always steam-tight, the action of the spring gives them a rotative motion, and this action has the effect of keeping the cylinders perfectly true with a minimum of friction, increasing thereby the power developed, and effecting a saving in the consumption of fuel. To fit these rings into the cylinder—(1) put the engine on the bottom centre, take the two half-rings (without the spring) and put them into the cylinder, see that they move freely from top to bottom, keeping the joints barely $\frac{1}{16}$ of an inch open in the tightest part of cylinder. (2) Bring the piston to the top, put the two half-rings in (still without the springs), placing four pieces of lead wire about $\frac{1}{4}$ of an inch long and $\frac{1}{16}$ of an inch thick on the top of them — then put the junk-ring on and screw it down tight — upon taking it off again there ought to be $\frac{1}{32}$ inch free. (3) Wipe dry the cylinder and outside of rings, put the spring into rings, placing clams on the inside, and screw down easy, place the hoop on outside of rings, screw the joint up close, then tighten the clams to keep rings from slipping back, take the hoop off and place the rings into the cylinder, take the clams off and put on the junk-ring. It is stated that over 10,000 of these pistons have already been fitted, which would indicate their superiority over the ordinary piston rings and springs. There are numerous other patent pistons and springs in the market, but as they all more or less resemble those already described engineers should have little difficulty in adjusting them, the main point to remember being that they should never be made too tight a fit in the cylinder.

CHAPTER III.

The Slide Valve—Importance of keeping it in Order—Arrangements for Reducing Friction—The Piston Valve—Its Advantages and Disadvantages—Haythorn's Patent Piston Valve—Radial Valve-Gears—Reason for their Adoption — Hackworth's, Marshall's, Bremme's, Kitson's, Kirk's, Bryce Douglas', Joy's, Walschaert's, Morton's, Brown's, and Sisson's Radial Valve-Gears — Requirements of a Theoretically Perfect Slide Valve-Gear—Comparison of the Different Valve-Gears.

In order to obtain the greatest efficiency and economy from the working of the engines it is essential that they should be periodically examined and maintained in the best possible order. One of the most important parts being the slide valve, a fair knowledge of its action, its proportions, the changes that can be made by the various alterations of lead, steam, and exhaust lap, the position of the sheaves, and the setting and adjustment of the valve-gear, is necessary—if efficiency and economy are to be studied. Too much attention cannot be bestowed upon the setting and adjustment of the valve, and, when tightening the nuts or cotters on the valve-spindle for the purpose of securing it, great care should be taken to prevent it being jammed. The valve should be perfectly free to move on the spindle, but without the slightest end play. This will admit of the steam pressing it against the cylinder face, and also ensure the minimum amount of friction. The amount of lead to be allowed at the top and bottom of a slide valve varies, of course, with the class of engine; but, generally speaking, with an increased speed of piston the lead may be increased.

Slide valves should be regularly and carefully examined, and whether the valve be a piston or the common slide all working surfaces should be perfectly steam-tight. The valve-faces should be kept as smooth as possible, and every opportunity taken to ease any hard or rough places that may appear, in order to reduce the friction to

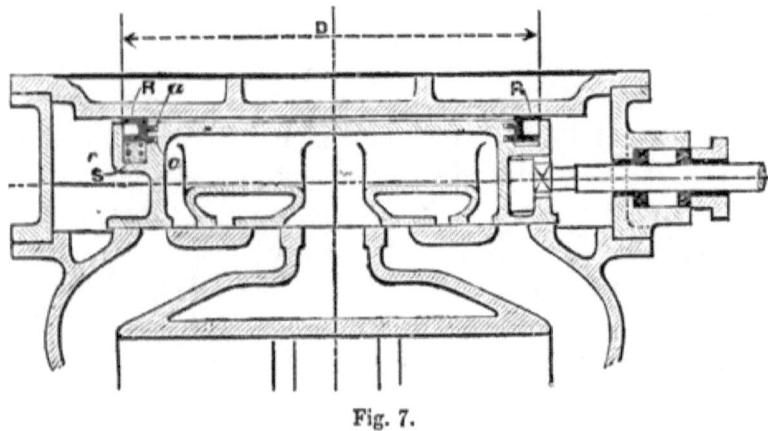

Fig. 7.

a minimum. One great objection to the use of the ordinary slide valve is the excessive resistance, which is almost wholly due to the friction of its working surfaces, and which increases when the pressure is increased. In order to relieve the pressure on the faces of large slide valves, such as are used in marine engines, the usual practice is to fit a ring or some similar contrivance on the back of the valve, so as to reduce the area exposed to the steam pressure, and thereby diminish the total pressure on back of valve.

One method, and perhaps the best, is that known as the "Church" equilibrium ring. Fig. 7 is a sectional elevation of a slide valve with this ring fitted on the back. Fig. 8 is a plan of the ring, R is the principal ring of channel section as shown. This is kept up against the planed face of the inside of the valve chest cover by means of springs, SS; there are also one or more Ramsbottom rings, $a\ a$, for the purpose of prevent-

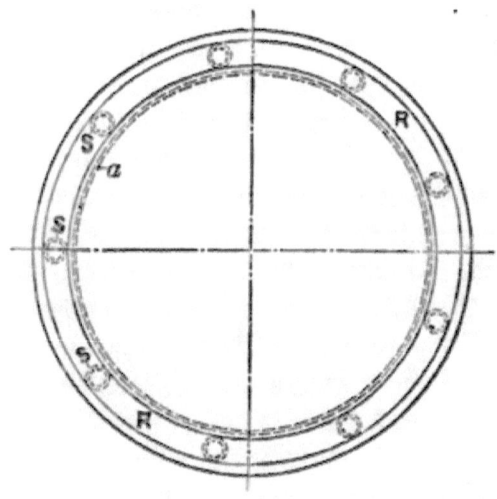

Fig. 8.

ing the escape of steam past the ring R and the groove inside which it works. The ring R is thus kept steam-tight on two faces—viz. its upper face and the inside face of the ring, thus preventing any escape of steam into the space on the back of the slide valve enclosed by R. By this means the steam pressure on the working face of the slide valve is reduced by the amount of pressure on an area equal to the area of the ring R on its outside diameter.

Numerous arrangements of this kind have from time to time been fitted to the back of slide valves for the purpose of reducing friction, but none of them have given complete satisfaction, indeed some actually produced more friction than they were designed to reduce, while others after working for a time got out of order and were then worse than useless. It was to get rid of these disadvantages that the piston valve was designed. It consists essentially of two pistons working in a cylindrical chamber and connected by a rod, the distance between the pistons being equal to the width of the exhaust port in an ordinary slide valve, while the depth of the pistons corresponds to the length of the valve-face on each side of the exhaust port. Steam is admitted outside the pistons, and it exhausts from the cylinder between them, from whence it goes into the exhaust passage in the usual manner, or *vice versâ*. The advantages of a piston valve are: the port area is about three times larger than that of a flat valve of the same dimensions across, and the steam pressure, being equal all round the valve, is balanced, the friction accordingly being greatly diminished. Great difficulty is said to have been experienced by some engineers in keeping these valves steam-tight, although others maintain they are as tight, perhaps tighter, than flat valves; but, notwithstanding this, they appear to be rapidly coming into favour and are now generally fitted in triple expansion engines in place of the ordinary slide valve.

When the valve is constructed so as to be drawn out by the top, steam is admitted between the valves, and the upper piston is made larger than the lower, so as to balance the weight of the valve; but when it has to be drawn downwards, the steam is admitted by the outer ends of the valve, and the lower one is then made the larger of the two.

Piston valves can, however, be made and kept quite steam-tight and practically frictionless, provided they are properly constructed and supplied with springs and efficient means for accurately expanding the ring radially, adjusting it with the utmost nicety, and securing it firmly when it fits the chamber. It should not, when thus constructed, be more liable to get out of order than a solid plug, and, in first cost, it will compare most favourably with any other kind of piston. For large cylinders two or three piston valves can be arranged instead of one large slide valve, and there are instances known where old low-pressure slide valves have been replaced by a casting containing two piston valves bolted on the cylinder face, which gave every satisfaction in working.

There is little doubt that the prejudice against the use of piston valves is mainly due to the fact that so many of them, fitted with complex spring arrangements, have been tried and found wanting. In many of these the rings collapsed under the compression of the steam at the turn of each stroke, thereby in a short time knocking themselves to pieces; and, further, the resistance to wear of the chamber at the steam ports being less than that at the solid portion, this allowed the ring to gradually produce enlargement, or "barrelling," at that particular place, thereby increasing the collapsing action.

This has resulted in many cases in these spring piston-rings being removed and solid piston-blocks put in their places. The advantage of this plan is that it prevents "barrelling," but the obvious difficulty is to keep them steam-tight after the slightest wear takes place, and consequently these solid pistons have to be frequently renewed. Perhaps the best method of keeping piston valves tight would be to employ the form of ring shown in Fig. 1. This ring being, when screwed up, practically solid, it possesses all the advantages of solid pistons, while by means of the joint the slightest possible amount of wear can be taken up by simply putting in a thicker liner, thus keeping the piston always a floating fit.

One of the most successful piston valves is that known as Haythorn's patent, which has been in use for over five years, and the patentees claim that in addition to the increased effective power, due to the diminished friction of the working parts, there is a great saving effected in repairs, in renewals, and in the disuse of lubricants.

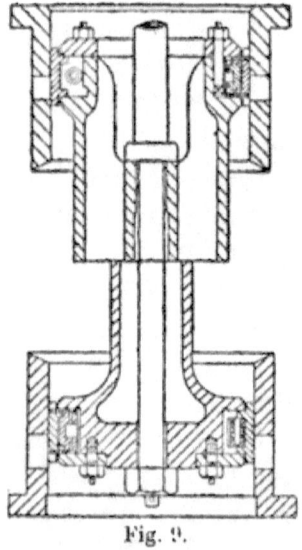

Fig. 9.

Fig. 9 shows a vertical section and plan of one of these piston valves, the lower half of section being arranged as a solid or close valve, and the upper one being hollow, and forming communication between top and bottom ends of valves. The arrangement of rings and adjustments is precisely similar to that used for small cylinders, the rings being adjusted to fit the valve chambers, and firmly secured; the outer ends, being prolonged, give both a free admission and exhaust. These valves can be readily arranged on the "trick" principle, and, from the fact of their being securely prevented from expanding, a multiple-ported chamber is unnecessary, only a sufficient number of ties to keep the chamber together being required, thus reducing the diameter necessary for providing the requisite port area.

As already stated, most builders now construct triple expansion engines with their valves either in front or at the back of, instead of between, the cylinders, and as the three cranks thereby only occupy about the same space as that previously required for two, it is evident that in many cases there will not be sufficient room on the shaft for three pairs of eccentric sheaves. Some modification of the old form of valve-gear, therefore, became necessary, and a number of gears principally designed to overcome this difficulty, and known as radial valve-gears, are now in use. Radial valve motions are very convenient, as they do away with the necessity of eccentrics, and thereby allow the bearings to be brought up close to the cranks, thus giving greater strength and stability, at the same time making the engines more compact, and they have been adopted in many instances, we believe, on this account alone. As these valve-gears are not generally known amongst marine engineers, a description of some of them which have been pronounced the most successful, and are now generally used, may be serviceable.

Figs. 10 and 11 show Hackworth's radial valve-gear. One end B of the rod BD is constrained to move in a circular path about the centre of the crank shaft, the other end D of this rod is constrained to move in the slot of a link D, the result being that any point, as E, in the rod will describe an ellipse, the direction of this

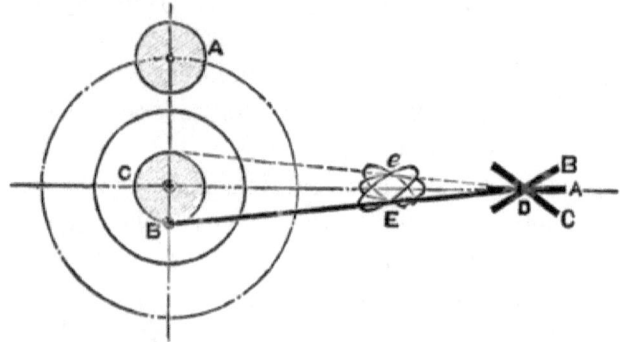

Fig. 10.

ellipse being regulated by the position of the link D. Fig. 10 shows the link D at the end of the rod BD, whereas Fig. 11 shows the link at an intermediate position between the point E, which describes an ellipse, and

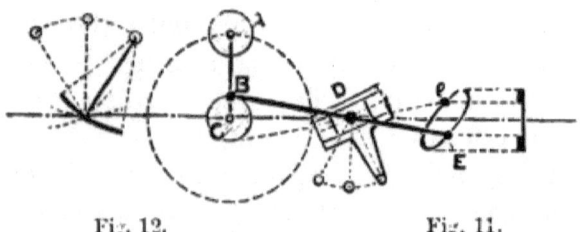

Fig. 12.   Fig. 11.

the point B, which describes the circular path. Sometimes the link is curved, as shown in Fig. 12. By moving the link D into different positions about its centre D as fulcrum, varying grades of expansion are obtained, and by this means also the engines are reversed. When the D is in a horizontal position, so that it lies in the direction of the straight line AC, the result is that no motion is imparted to the slide valve; but when the link is in any other position, such as B or C (Fig. 10), the valve is in gear for going ahead or astern.

The great advantage of this gear, in addition to doing away with the eccentric, is the improved motion imparted to the valve, there being two quick and two slow motions in a revolution, the quick motions occurring at the point of cut off and the slow motions during exhaust and previous to the admission of steam. A considerable variation in the amount of cut off is possible with this arrangement without wire drawing from a small opening and slow closing of the port, as is the case with the common link motion. The chief objections urged against this gear are the excessive friction, and consequent wear on the sliding blocks, and the liability of so many pins to derangement. The first of these is the most valid, and it has been overcome to a great extent by fitting rollers instead of blocks. In both ways, however, this gear has worked fairly well; and for engines of small power it is a very convenient arrangement, especially when much variation in the amount of cut off is required.

Marshall's gear is a modification of Hackworth's, and differs from it in the method of getting the oblique motion of the rod end.

Figs. 13 and 14 show the plan adopted by Mr. Marshall. Here the eccentric rod is hung by means of a rod from the end of a lever on a reversing shaft, in such a way that it moves on the arc of a circle inclined to the centre line. The motion is not quite so perfect as with the inclined sliding bar, and necessitates double ports to the bottom end of the slide valve, in order to get as much opening to steam as there is at the top end; but there is less friction, and, on the whole, it works most satisfactorily. The pins require to be of good size, and they should all have adjustable brasses to provide for the large amount of wear which of necessity comes on them.

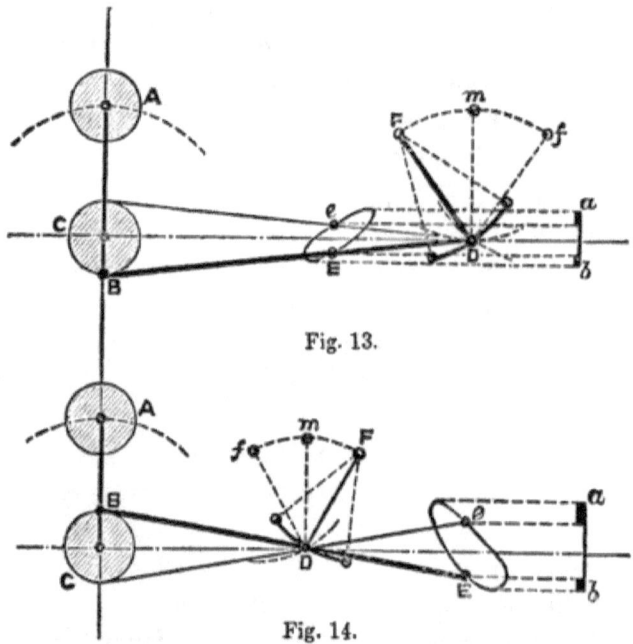

Fig. 13.

Fig. 14.

Fig. 15 shows Bremme's gear, which, it will be seen, bears a striking resemblance to Hackworth's gear, already described; indeed, they are so much alike, that further description will be unnecessary.

Fig. 16 shows Kitson's gear. In this gear the circular motion of the crank pin A imparts a reciprocating motion to the end S of the rocking lever (*n r s*), which moves about (r) as fulcrum, the other end (n) giving an oscillating motion to the bell-crank slotted link (b L m), which works about D as centre, the point b of the radius rod (cfb) works in the slot of this link, and according to its

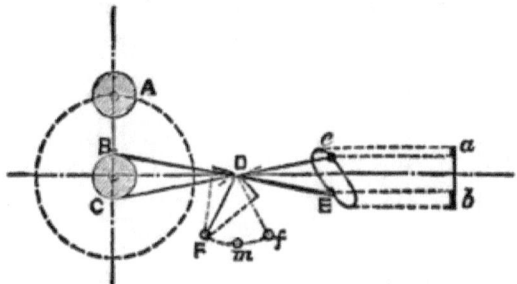

Fig. 15.

position determines the motion of the lever (fcd) which works about o as fulcrum, the end d being attached to the valve-spindle, and the end f attached to the crosshead pin D by means of the connecting link f D.

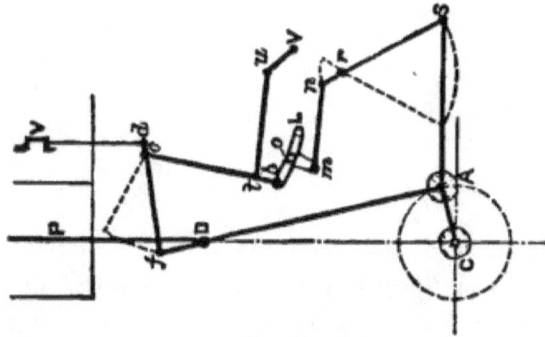

Fig. 16.

Fig. 17 shows Kirk's gear. The motion of the valve V is obtained from the piston-rod crosshead, through a bell-crank link L. This being placed a proportionate distance from the centre or fulcrum f of the air-pump

32   Triple and Quadruple Expansion Engines

lever gives a motion (when in mid gear) to the valve V, equal in extent to the "*lap + lead*" from the crosshead, the extent or port opening being obtained from the vibration of the bell-crank link L, which is driven from the connecting rod by means of a link M connected to a compensating lever K, by means of which latter the error due to the working of the connecting rod is corrected.

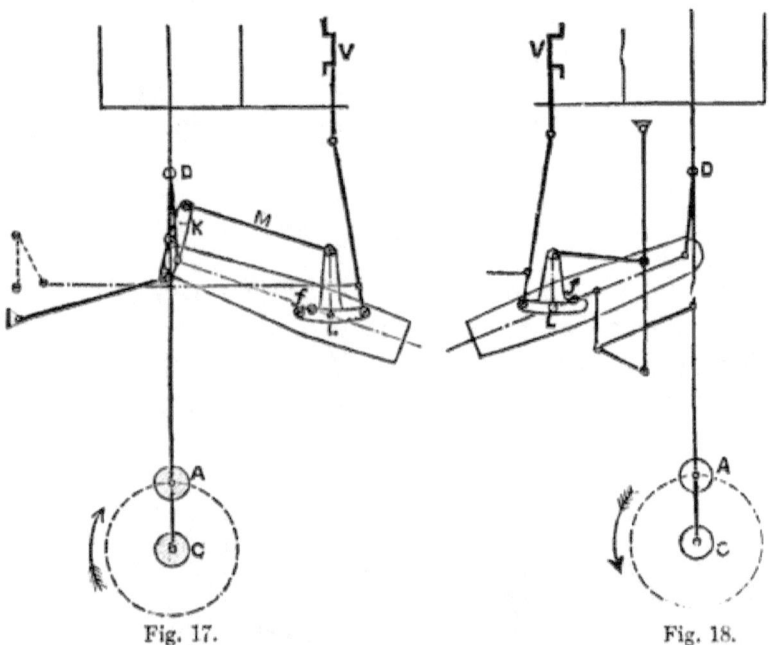

Fig. 17.   Fig. 18.

The advantages of this gear are that, by dispensing with eccentrics, the engine is somewhat shortened in the ship, and access to the main bearings and connecting rods is much facilitated. It is only applicable where the valves are at the back of the cylinders.

Fig. 18 shows Bryce Douglas' gear, in which a bell-crank link is also used. It bears a close resemblance to Kirk's gear, both using the motion of the air-pump lever.

Fig. 19 shows Joy's valve-gear. PQ is the connecting rod, B is the piston-rod, CP is the crank. Take any point R in PQ. This point R will describe an elliptical figure as shown by the dotted lines. RT is the compensating lever; this is a very important detail in this gear, since the point S, in the compensating link RT, describes a flattened ellipse as shown in dotted lines, thus causing the fulcrum F to have an equal and symmetrical motion. While S describes the flattened ellipse before mentioned, the fulcrum F slides to and fro in the slotted link HH, fulcrumed at F, at the same time the point E on SF produced describes an elliptical figure as shown by the dotted lines. This determines the motion of the slide valve, which is obtained through the connecting rod or link ED, and so to the valve rod A. Draw horizontal lines from the top and bottom of the ellipse described by the point E, and where they intersect the vertical line ab, determine the amount of travel $= ab = t$. As in Hackworth's gear, there are two quick and two slow motions in a revolution, and just when required the amount of opening is equal at both ends of the valve, and an earlier cut-off can be obtained without an excessive amount of lead and compression, or too early opening of exhaust. This gear is now one of the most successful of its kind, and is extensively adopted by marine engineers. The chief objections are that it prevents the principal moving parts being readily got at when working, and further, a very small amount of wear at the joints would alter the motion of the valve, and cause the gear to clatter badly. This may be remedied, however,

34   *Triple and Quadruple Expansion Engines*

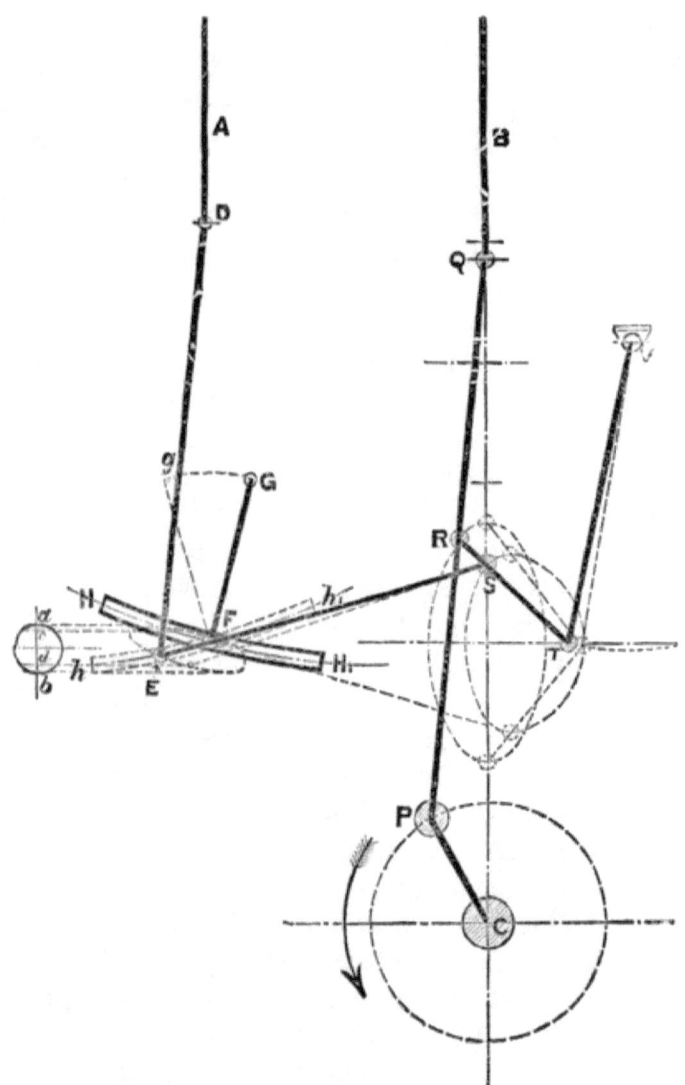

Fig. 19.

by making the pins larger, and by having adjustable joints.

Fig. 20 shows Walschaert's valve-gear as applied to a marine engine. C is the crank shaft, A is the crank centre of the crank shaft. This is virtually an eccentric, the crank itself is reflected back as shown, so that point B describes a small circular path round the pin, but no eccentric is used. The link L is a fixed link, and is fulcrumed about O as a centre. L derives its motion from the point B, to which is attached the rod B b. In L works or slides a block to which is attached the radius rod c b. The point c is attached to the point c in the long lever d c e; one end of this long lever, viz. d, being attached to the valve-spindle, which is guided in a fixed bracket, the other end of the valve-spindle being secured to the slide valve V. The end e of the long lever is attached by means of link e f to the crosshead pin D.

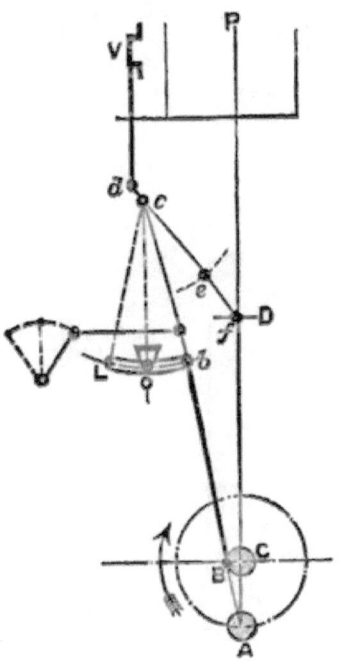

Fig. 20.

It will be seen that the slide valve V derives its motion from the resultant of the two motions, viz. of the point c, as derived from the motion of the link L, according to the position of the sliding block b, and the motion due to the lever d c e rocking about c as a centre, the motion of e being equal in extent to the travel of the crosshead. In this gear the piston-rod connection moves the valve through its "*lap + lead*," whilst the motion of the point B gives the port opening. In order to reverse the engines, or to obtain expansion, the sliding block *b* is moved in the link L. When *b* is in the centre of the link, at *o*, there is no motion of *c*, but, by shifting *b*, either to right or left of *o*, the engines will go ahead or astern.

Fig. 21 shows Morton's valve-gear, in which reversion and expansion are obtained by sliding the valve-spindle connection across a vibrating bell-crank link, whose neutral axis is fixed. Thus, when the rod H is in a line with the valve-spindle, the gear is in neutral position, the valve movement being only "*lap + lead*," but by moving the top end of H along the quadrant, the extra extent of travel required for the port opening is obtained, and according as H is on one side or other of the centre of quadrant, so will the valve be set for going ahead or astern. This gear corrects more perfectly the error arising from working off the connecting rod than either Joy's or Brown's gear.

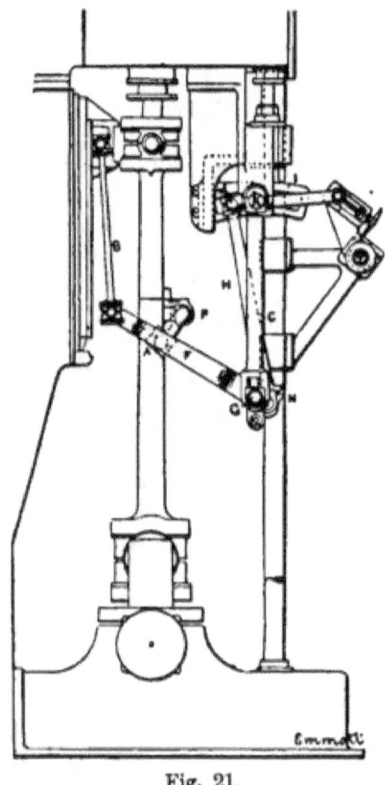

Fig. 21.

Fig. 22 shows Brown's gear, in which the link M is connected to the point *e* in the connecting rod, the point *f* being attached to the valve-spindle by means of a connecting rod *d f*. The point *g* is attached to the end of a rod *g* G, which is capable of sliding and swivelling in the centre G; the point *g* will thus describe an angular line, slightly curved, giving the necessary motion to the point *f*, which determines the travel of the slide valve V.

Fig. 23 is an illustration of a radial valve-gear invented by Sisson. It resembles Joy's gear wrought by Kirk's connecting-rod motion, and is considered a very good gear.

A new arrangement of valve-gear, for which several important advantages are claimed, besides simplicity of construction and economy in first cost, has been quite lately invented by Mr. James Howden, of Glasgow, through whose courtesy the author is enabled to give two illustrations of it, as seen in Figs. 24 and 25. In this arrangement there is an independent counter-shaft placed in a convenient position parallel to the crank shaft, from

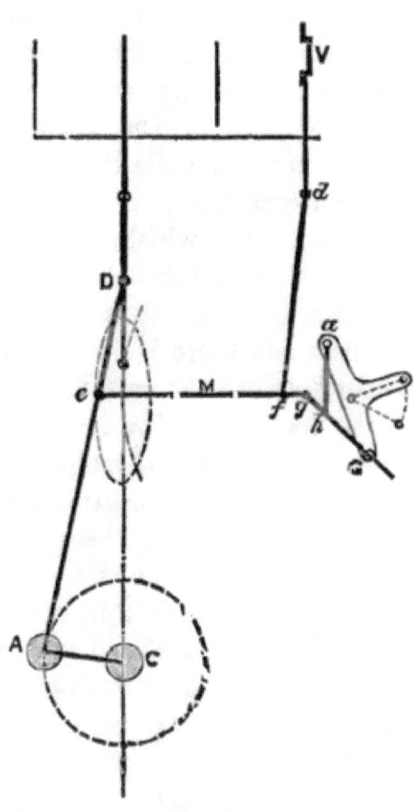

Fig. 22.

which it may be driven either by spur-gearing connecting rods as shown, or by any other suitable means. This counter-shaft may be termed the valve shaft, as upon it are placed the eccentrics which give the requisite travel to the valves. Each valve-spindle has only a single eccentric provided for it, which is so fitted that it is capable of sliding across upon parallel surfaces on opposite sides of the shaft for regulating the cut-off and for reversing the engines.

Each eccentric is made in halves, and has fixed to it two steel bars, which, when the eccentric is closed on the shaft, enter through diametrically opposite slots into the interior of the shaft, which is hollow. In the interior of the shaft there is an inclined steel bar which fits and works between the suitably shaped inner ends of the bars of the eccentric. The inclined bar rotates with the shaft, but can be moved longitudinally in it, and when so moved shifts the eccentric across in one direction or the other accordingly. The counter-shaft is, by preference, made with sections for each valve, simply coupled together, and capable of easy separation for examination or repair. The inclined bars are all connected together by intermediate rods or shafts and couplings, also admitting of easy separation; and the longitudinal movement for adjusting the eccentrics is imparted at one end by means of a sliding bar connected by a swivel coupling. The sliding bar may be moved by hand gear, or by means of steam, or hydraulic motor apparatus controlled by hand levers.

Having described most of the principal radial valve-gears now in use, we would observe that in a theoretically perfect slide valve-gear the motion would require to be a series of jerks to give a quick and full steam admission, then a period of rest, next a quick cut-off, then another period of rest, and finally a quick exhaust opening ; and all this should be accomplished with the same gear with which expansion and reversion are effected. It is evident, therefore, that the form of valve-gear which most nearly approaches the above requirements is the best. All forms of valve-gear, including the ordinary link motion, whether considered as steam - distributors or mechanical arrangements, have their merits and demerits; but the consideration of these being beyond the scope of this work will not be touched upon. In looking at these different gears, merely as mechanical arrangements, the first thing a practical engineer takes into consideration is the number of moving wearing parts and the strains upon them. Viewed in this light, the ordinary link motion, in our opinion, compares favourably with most of the other gears, and in spite of the opposition it has met with it will doubtless be long before it is entirely displaced. The most likely thing

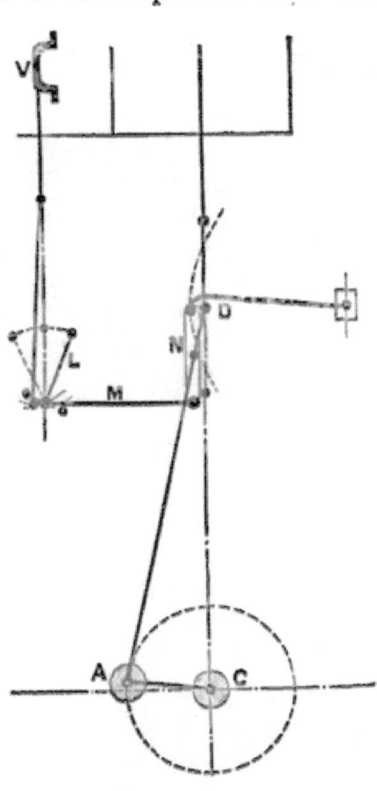

Fig. 23.

to hasten this would be the endeavour of engineers to so contract the fore and aft length of triple expan-

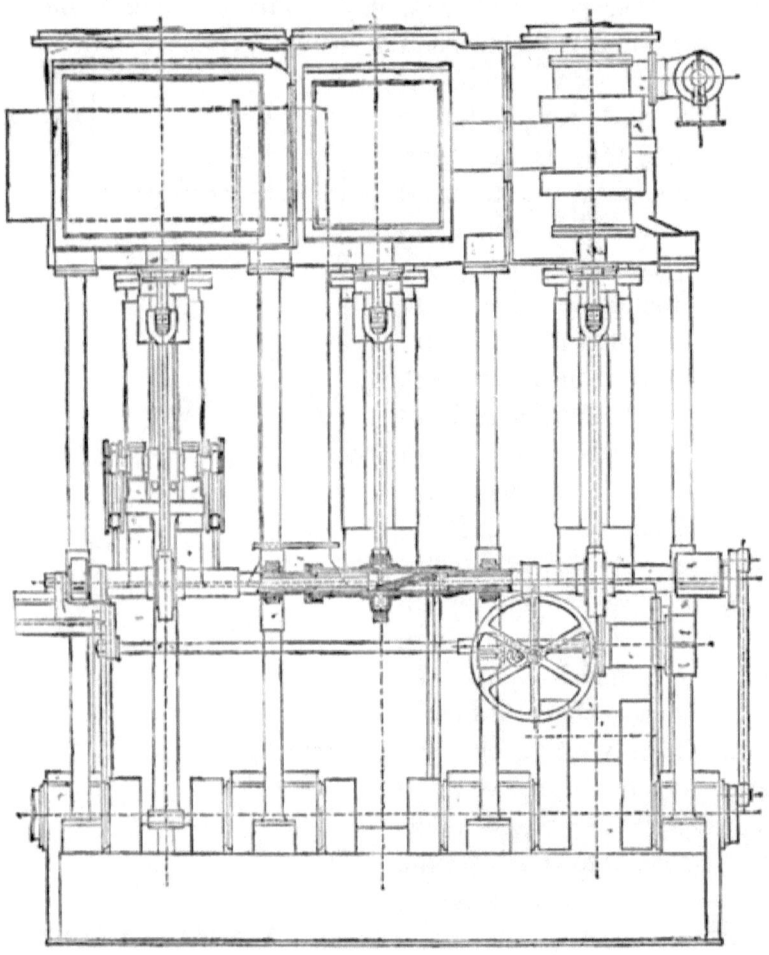

Fig. 24.

sion engines that they would not take up more space than that previously occupied by the two-cylinder compound engines, and this saving in space alone will generally more than repay the extra cost of construction and royalties, which handicap most of the gears when compared with the ordinary link motion. One serious objection, however, to the use of the link motion for triple expansion engines is that it cannot be readily adapted to the altered position of the valve chests, unless vibrating levers are introduced, and these add so much to the cost that most of the radial valve-gears would then compare favourably as to price, besides being better suited to such altered conditions. This is, no doubt, true to a certain extent in the majority of instances, and the introduction of vibrating levers is an evil that is to be avoided wherever possible, more especially in connection with the link motion, which depends so much upon its comparative fewness of joints and their careful adjustment for its well-known efficiency, and its great convenience in facilitating the expansive use of steam within certain defined limits; but it will be seen that this serious objection does not apply to Howden's arrangement, as the introduction by him of a counter-shaft gives the same result on the valves as if the eccentrics were actually placed on the crank shaft itself. While speaking favourably of the link motion, it is only fair to state that some of the other valve-gears which we have described can be constructed at a less cost. Bearing this in mind, as well as the fact that these same gears are better steam-distributors, it may be interesting to complete the comparison by referring to the cost of maintaining them in working order. Take Hackworth's gear, for instance, which has a large eccentric and sliding bars, and which has more sliding surface than other radial gears, the testimony of those who have recently adopted it being to the effect that it requires fewer repairs and less adjustment than the link motion;

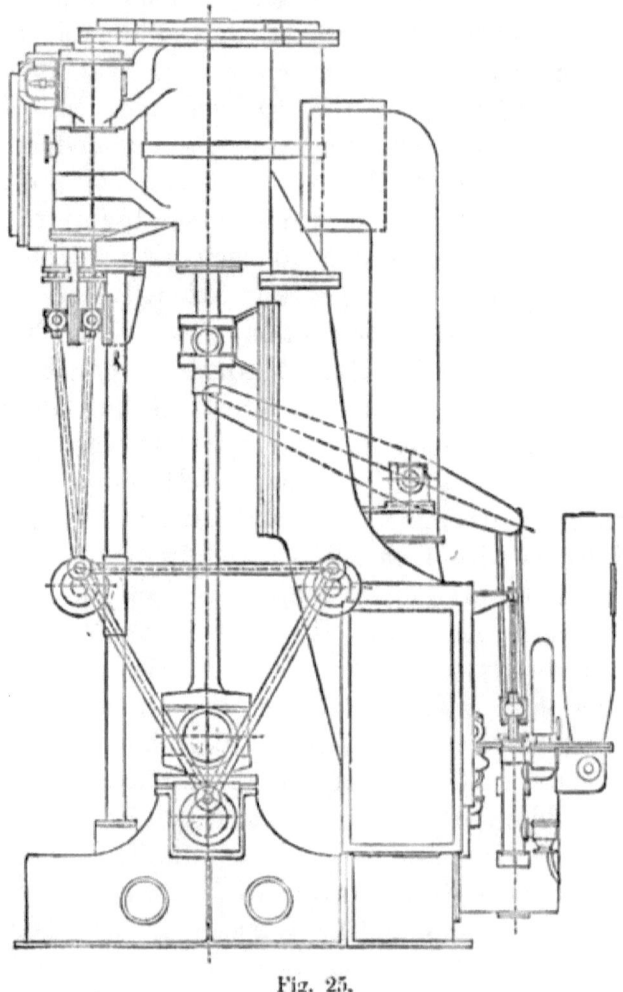

Fig. 25.

and report speaks even more favourably in this respect of Marshall's gear. These gears have only five working wearing parts when the engines are running, as compared with ten in the direct acting link motion, whilst Joy's, Walschaert's, and Morton's have about the same number as the link motion, and yet are said to wear better. Some of the other valve motions have more than that number of working parts, and where so many are necessary to attain the desired result, we hardly think the end justifies the means.

CHAPTER IV.

The Sequence of Cranks—Greater Economy with the High-Pressure Crank Leading—Proportions for Cylinder Diameters—Three Cranks better than Two—The Steam-Jacket—Its Advantages or otherwise—Most Economical on Intermediate and Low-Pressure Cylinders—Friction of Engines—Experiments thereon—How to Reduce it—Packing of Glands — Metallic · Packings — Duval's — United States — Macbeth's Patent—The Ideal Packing.

THE sequence of the cranks—that is, the order in which the cranks should follow each other—is at present an undecided question. Sometimes the sequence recommended was, low, intermediate, high; commonly known as low-pressure leading. This order admitted of smaller receivers than with the high-pressure crank leading, or the ordinary sequence of high, intermediate, low. In this latter order the three events—admission to the following cylinder, compression in one end, and release from the other end of the exhausting cylinder—were nearly simultaneous. The advantage of the ordinary sequence has been confirmed by the greater economy obtained from engines with the HP crank leading than from similar engines with the LP leading. Three causes contribute to this result—viz. first, the prevention of loss from a sudden fall of temperature in the released steam exhausting into the receiver; second, the following cylinder being saved from part of the bad effect of internal condensation; and, third, the turning forces being found to be much more

Triple and Quadruple Expansion Engines. 45

equable. It is considered by some authorities that an advantage arises from allowing the three separate ranges of temperature in the successive cylinders to overlap one another instead of forming the minimum subdivision of the total range; and the range in the LP cylinder might be considerably greater than in either of the other two. The equality of initial stresses also affects the strength of parts, and the extent of bearing surfaces and the equality of horse-powers influence the ease and regularity of turning.

When there are three cranks and a high boiler pressure a very good proportion for the cylinder diameters is 3, 5, and 8, and it must be kept in mind that three cranks are always better than two, and that a three-crank engine fitted with one of the modern forms of valve-gear occupies only the same length as a two-crank compound with the ordinary link motion.

Regarding the vexed question as to what extent and in what manner the steam-jacket is advantageous or otherwise, very little additional knowledge appears to have been gained upon the subject since the introduction of Mr. Cowper's well-known hot pot or intermediate heater; in fact, those who are considered the best authorities disagree even where they themselves have had large practical experience of it. It is sometimes advised to jacket only the high-pressure cylinder; sometimes only the low pressure, and sometimes to jacket all three. Experience would seem to indicate that, in order to secure the maximum efficiency, assuming the jacket to be efficient, the best system would be the latter—that is, to jacket all three; and that since the waste of the engine is measured by the waste of its most wasteful member, to omit the jacket from any one cylinder insures that the total loss of heat in the whole engine will be increased by just the amount by which the waste is increased in that one cylinder.

Mr. J. Wright, who in the course of a number of years had taken about 170 sets of indicator diagrams from various cylinders with the ordinary steam-jacket, both on and off, and others with the inter-heater he had invented, either on or off, hoped thereby to establish some rule. The experience gained by him, however, showed that about fourteen out of every fifteen of the cylinders experimented upon were overheated by their steam-jackets; indeed, some so much so that they entailed a positive loss on the working of the engine. Whenever the temperature of the steam in the cylinders was only slightly increased by the jacket there appeared to be considerable economy, but when the temperature was increased largely the economy partly or entirely disappeared. This appeared to be in consequence of the enormous friction produced on the engines by superheated steam. Although the diagram showed a large increase of power in the cylinder, the engine lost speed when working against the same load. Mr. Wright has introduced in his inter-heater an automatic arrangement for controlling, within certain limits, the temperature of the steam in the jacket, so as to vary with the main pressure in the cylinder. This had not hitherto led to any good results; but without some such controlling influence, Mr. Wright considered, speaking from his own experience, that steam-jackets on cylinders were rather a disadvantage than otherwise.

Although there is such a great difference of opinion regarding the advantage, or otherwise, we consider the weight of evidence is in favour of their use, at least on the intermediate and low-pressure cylinders, and the following are some of the reasons adduced why jackets would be more usefully applied to these cylinders than to the high-pressure one: 1. The area of cylinder end exposed to condensation is less in proportion in the high pressure. 2. Condensation in the high pressure lubricates the cylinder, and this dispenses with oil and its consequent evils at high temperature. 3. The steam gets wetter and wetter as it passes through the cylinders. 4. Condensation in the low pressure produces more ultimate loss than in the high. 5. A steam-jacket on the high pressure would be partly idle during admission, and considering the high mean temperature in that cylinder it would be very little better than efficient lagging. 6. A steam-jacket would be very effective on the second and third cylinders, on account of the large gradient of temperature between the boiler steam and the steam in these cylinders. Probably, on account of this diversity of opinion, engine-builders are reluctant to construct engines with steam-jackets, and consequently a large number of engines are made without them; but, in the case of those that have them fitted, the best results would probably be obtained, as stated above, by using the jackets on the intermediate and low-pressure cylinders only, and with a steam pressure not more than 10 lbs. above the mean pressure in each cylinder. It is very likely that in some cases where the jackets have not been efficient it has been due to the excessive pressure of the steam admitted into them. This insures the temperature of the cylinder being maintained nearly stationary; the drain cocks should be kept slightly open to allow the jackets to be kept clear of water.

The friction of engines is a subject on which very little light has been thrown until lately, when Professor R. C. Carpenter, and Mr. G. B. Preston, of America, made experiments with the object of ascertaining the distribution of internal friction. The result of these experiments was described by Professor Thurston in a paper read before the American Society of Mechanical Engineers, to whom we are indebted for the information now given. It becomes an interesting and vitally important problem to determine just how the friction of the engine is distributed among its various moving parts, its journals and guides, stuffing-boxes and piston-rings. This has hitherto been regarded as a problem incapable of solution, since it was presumed that the total friction and the friction of the various parts of the engine would be so extremely variable with the alteration of the load that it would be impossible to measure the friction of the machine, part by part, and to sum up the whole correctly. It having been found, however, that the friction of the engine is invariable in any measurable degree with the variation of power, it becomes possible to analyse the engine with its various friction-producing elements, and measuring up each element of friction by itself, sum up all for the total. The result of the experiments showed conclusively that the friction was substantially the same with no load at all as it was with all loads up to its full power. It was further discovered that the most important item of friction waste in every instance is that of lost energy at the main bearings. In every case it amounts from one-third to one-half of all the friction resistance of the engine. The second highest item is, in all cases apparently, the friction of the piston and rod, the rubbing of rings, and the friction of the rod-packing. This is a very irregular item, as would naturally have been anticipated, and amounts to 20 per cent. and upwards. The third item in order of importance is the friction of the slide valve, in the case of engines having unbalanced valves. This is seen to be hardly a less serious amount than the friction of the shaft and of the piston. But it is further seen at once that this is an item which may be reduced to a very small amount by good design, as is evidenced by the fact that it has been brought down from 26 to 2·5 per cent. by skilful balancing. Ninety per cent., therefore, of the friction of the unbalanced valve is avoidable or remediable, and this has now to a great extent been duly attended to in modern marine engines by the adoption of the piston and balanced slide valves. The importance of removing this friction is readily perceived when it is considered that not only is it a serious cause of lost work, but also of wasted power and fuel. The frictions of the crank pin of crossheads, and of eccentrics, are the minor items of this account; but they are comparatively unimportant.

We are now, for the first time in the history of the theory of the steam engine, in a position to say just where the losses of the machine occur in detail—how we are to endeavour to reduce them, in what degree we may hope for such gain, and where it is to be found if effected at all. The first and most remarkable fact to be noted is the extraordinary amount, absolutely and relatively, of the friction of the crank shaft. This amounts to nearly one-half of the whole waste, and to from 5 to 10 per cent. of the whole power of the engine.

Here is evidently the first place in which to seek further improvement. If this item can be brought down as low as it generally is in car axle journals, the efficiency of the engine as a machine will be increased by about 10 per cent. in ordinary engines, and by about 5 per cent. in the very best cases. How this is to be done can be best determined when we find out the causes of so extraordinary and previously unsuspected a loss. The only conditions that are apparently accountable for this waste are the continuous rotation in one direction, and the uninterrupted pressure of the journal in its bearing, aggravated probably by imperfect lubrication. Could the oil bath system in method and in results be always secured here it would be only reasonable to expect that the friction might be enormously reduced. It would even, in many cases, if not in all, pay well to have a thoroughly reliable system of lubrication by means of a forcing-pump, that would insure the support of the journal upon a cushion of lubricant, thus making its action analogous to that of the "gliding railway" of Giffard, and the "water bearing" of Shaw and of others.

The second and most obvious conclusion is that the slide valve should be so balanced and connected as to cause the least possible waste by friction through its motion or that of its moving connections; and there is no probable line of improvement so certain to yield a large and profitable result as this. No engine can be considered as efficient which is not either provided with a balanced valve or a system of valve-gear in which the loss in this direction is rendered insignificant; and herein lies an opportunity for increasing the efficiency of ordinary engines at least 5 per cent., and of the best of engines with unbalanced valves 2 or 3 per cent. It is better, in many cases, to have a valve which is balanced, though slightly leaky at times, than to use an unbalanced valve, though absolutely tight at all times; and the simple fact here revealed, that nine-tenths of this friction may be easily avoided, is most important.

The third item in order of importance is the friction of the piston and piston-rod. This is as great as that just referred to, and is vastly more variable, according to the class of engine, and probably even in the same engines, with differences in their handling, and especially in setting up the packing and springs. The metallic-packings and the unpacked pistons and rods now coming into use will unquestionably do much to remedy this defect of the average engine. Meantime, with the older designs, it is perfectly possible to keep the piston and stuffing-box tight without wasting much power, or without slowing down the engines by the conversion of heat into work at points where the operation is likely to produce serious harm as well as considerable waste. Rings are much oftener too tight than too loose, and a stuffing-box should only be set up when the engine is running, and then only with fresh packing, and not more than is sufficient to check any visible leakage. New packing in a well-made box seldom requires much compression, and when it does become necessary to screw it down hard it is time to replace it by new. Any packing that requires severe compression when new should be promptly rejected. Some years ago numerous varieties of patent packings, similar to the original "Tucks," were greatly used, but as the working steam pressures have since then been considerably raised by the introduction of the triple expansion engine, it has become necessary to adopt something more lasting, and that can better resist the action of increased temperatures. Consequently asbestos packing of various kinds has been, and is still, much used. Some engineers object to the use of this mineral-packing on the ground that it unduly wears the piston-rods, while others dislike the screwing-up process altogether, and prefer a packing which produces steam-tightness other than by end-on compression. For this reason they prefer hard or metallic-packing, discarding the soft kind of packings altogether. Among the many varieties of metallic-packing, "Duval's" may be mentioned, which consists of fine brass wire plaited or braided into a square rope. It is cut into lengths and inserted in the stuffing-box in the usual manner, requiring no alteration upon the ordinary forms of stuffing-box. As soon as the packing becomes warm the expansion of the wire of which it is formed presses it firmly against the piston-rod and the interior of the stuffing-box, while the elasticity and extreme flexibility of the material insures that the pressure shall always be of a gentle and yielding character, although quite sufficient to insure a tight joint. The piston-rod rapidly acquires a smooth surface, which it subsequently retains. It does not in any way suffer from the heat of the steam, however high the pressure may be, and it cannot be said to show any sign of wear. It is said to give most satisfactory results, especially in triple and quadruple expansion engines, and one packing of the gland suffices for several long voyages without requiring any attention.

Figs. 26 and 27 show two forms of metallic-packing,

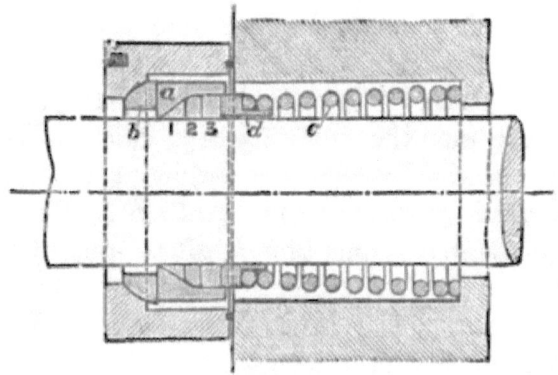

Fig. 26.

manufactured by the United States Metallic-Packing Company, and known as the U.S. Metallic-Packing. It consists of three babbit metal rings numbered 1, 2,

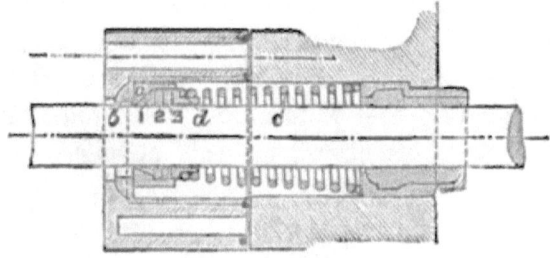

Fig. 27.

and 3, each having a small portion cut out to allow them to tightly clasp the rod; these three rings fit into a vibrating cup (*a*), which presses against a ball-and-socket-joint (*b*); this joint and the one between the vibrating-cup and the socket-ring are both carefully ground so as to be steam-tight in any position. The babbit rings are held in place by a spiral-spring (c), which presses a distance piece (d) against them; this piece, by being inserted a small distance into the vibrating-cup, is held in its proper position, and supports the smaller end of the spring. The pressure of the steam forces the babbit rings into the vibrating-cup, which is bored out conically, making them steam-tight against the rod. The principal feature of this packing is, that the two ground-joints, before mentioned, allow great freedom of movement, without impairing in any way the efficiency of the packing or its steam-tight quality. The results obtained from this form of packing are said to be very good, but they depend upon the quality and suitability of the babbit metal used. Some of these rings run considerably over 100,000 miles without repair or even inspection, and the cost of the renewal is trifling, but probably the greatest saving is effected by this packing prolonging the wear of the piston-rod and valve-spindles.

Fig. 26 shows the arrangement suitable for a piston-rod, and Fig. 27 one suitable for valve-rods. When using this packing the piston and valve-rods require to be turned up perfectly true and polished as fine as possible before the packing is applied, but when this is done the rods become as smooth and polished as a mirror after a few months' service, and remain practically of the same diameter and perfectly round section for several years, the amount worn off being less than $\frac{1}{64}$ of an inch for 100,000 miles run. The rods never become grooved unless something unusual happens, and the annoyance due to escaping steam is never experienced.

Good oil-cups require to be provided for each rod, and if this is done little or no attention is required beyond filling and attending to the oil-cups.

Fig. 28 shows another form of metallic-packing, manufactured by the U.S. Metallic-Packing Company, which is better suited for large-sized engines. It consists of eight packing blocks of anti-friction metal (*aa*), held in a stout ring fitted with springs (*bb*), which press the blocks up against the piston-rod PP. The pressure is sufficient to make a steam-tight joint, but not to cause binding of the rod. A ball-joint (B) is fitted at the top of the stuffing-box as shown, and is kept up to its bearing by a follower, fitted with springs (viz. R), and which is placed behind the packing; this allows of free play under all conditions. The whole is enclosed in a strong case

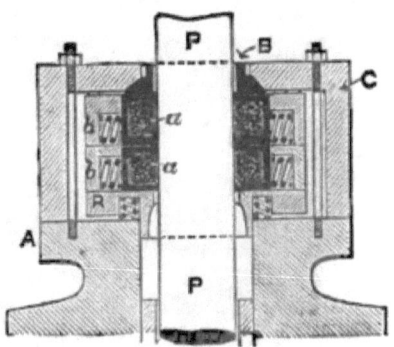

Fig. 28.

(C), which is bolted on to the cylinder cover. It has been found that the wear on the composition blocks (*aa*), Fig. 28, was only $\frac{3}{64}$ of an inch, after the engines to which this packing was fitted had run 24,000 miles, and there are many cases in which the packing had been run 150,000 miles without repairs.

Another metallic-packing which is said to be very suitable for rods which have become scored, worn oval, or even slightly bent, is that known as Macbeth's Patent, of which the following is a description.

A is a cast-iron box, Fig. 29, having a lid B containing

the universal stuffing-box. C is the cylindrical body of the stuffing-box, having gland pieces D and E. The body C and inner gland piece E are provided with coherical faces in contact with the joint rings FF. The joint rings FF are free to slide laterally upon the straight facings of the containing box A B, thus giving freedom for movement in all directions. Both the flat and the

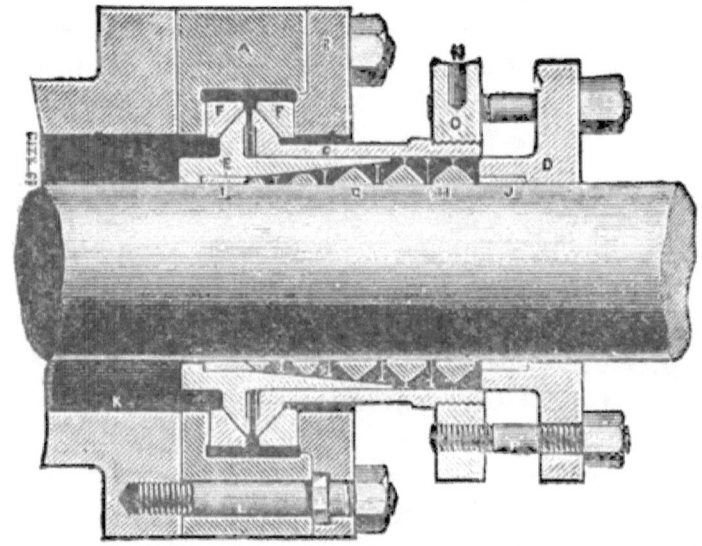

Fig. 29.

spherical faces of the joint rings are ground steam-tight. The packing G is inserted between the gland pieces D and E, and consists of rings of babbit metal, which are formed so as to be self-stripping at their ends, and which are backed up with asbestos packing H. The bolts P serve to compress the packing and keep it tight against the rod, at the same time putting the necessary pressure, in opposite directions, upon the joint rings FF, keeping them in contact upon the ground facings. N are holes provided in the flange of the body of the box, for turning the stuffing-box round the piston-rod, and causing the joint rings FF to rotate.

Amongst the advantages claimed for this invention may be enumerated the following: 1. The rod is kept perfectly steam-tight, even if the piston and cylinder have worn out of truth. 2. It works equally well if the crosshead and cylinder are not in line, or if the piston-rod has become bent from heating or other causes 3. No adjustment is required but what takes place from the outside; and by constantly rotating, the piston-rods are kept in beautiful order, and do not become scored nor oval. 4. There are no internal springs or complicated parts; for its construction is so simple that any unskilled workman can understand it, and it does not get out of order from grit or dirt. 5. It can be taken to pieces or put together again in a few minutes. 6. The passage of steam or air is perfectly prevented. 7. It can be applied to any existing piston-rod or cylinder cover, and saves time, labour, and material for packing, as well as the boring of cylinders and the fitting of new pistons. 8. It reduces friction, and increases the effective power of the engine.

The following instructions for fixing and working are given by the makers: Bolt the cast-iron box A to the flange on cylinder cover K, making the joint with red-lead cement. See that the split rings I and J (or D^1 E^1) are properly in their places. Put the brass box C and the button piece E together, and the rings FF upon them (see that the surfaces of the rings are well cleaned and oiled), then slide the stuffing-box into its place, and make the joint of the cover B with red-lead. To pack the box, first insert next to the ring I a ring of round asbestos packing, then the smallest set of babbit metal rings; then insert alternately two coils of asbestos and the next smallest babbit ring, and so on, as shown on drawing, until the box is full, taking care to place the asbestos H and the metal rings G so that the joint spaces between the ends do not come together. At the outside or gland end D of the box, place two coils of asbestos, and insert a small piece of asbestos in each of the joint spaces of the last babbit ring.

The asbestos rings will become compressed considerably after working a short time; when this has taken place sufficiently to allow of it, insert another babbit ring and two coils of asbestos.

When the babbit rings have become well bedded to the rod, and the asbestos fully compressed, very little screwing up of the gland will be required for a long time.

On account of the formation of the ends of the blocks of babbit metal, the rings will allow of their being completely worn away without being shortened.

If the boxes do not of themselves turn round the piston-rods during work, they should be turned occasionally by inserting a spike in the holes N. Until the babbit rings have become properly bedded, the rod should be slightly lubricated.

It is preferable to have the piston-rod skimmed up true, unless an ordinary packing material is used in place of the babbit metal rings, in which case, if the rod be in moderately good condition, skimming up may not be necessary.

There may also be mentioned a further variety of packing known as Nixon's, which possesses some entirely new features, and appears to meet, in a large measure, the requirements of a packing for high-pressure steam cylinders.

It consists essentially of a brass casing, which encloses the wearing-ring; this the makers prefer to be of the best gun-metal; it is divided into three segments, the joint being scarfed, and in addition is fitted with projections which make up a diameter equal to that of the casing. Attached to these projections are three springs, two of which are in tension, and the third in compression, the first two being adjusted by means of a single nut. One other point remains to be mentioned. The top of the casing has a knife-edge which forms a steam-tight joint against the neck bush; so far as the writer can learn, this forms an entirely new feature which is not found in any other patent packing. The packing is free to follow the rod whichever way the latter goes, and hence its name "Floating."

Although only in use about two years it is said to have already found considerable favour amongst marine engineers, and by those who have tried it has been found suitable and satisfactory for old and new piston-rods valve-spindles, and feed-pumps.

In addition to the foregoing there are many other varieties of metallic-packing now in use, but those already described give a very good idea of their general construction, and it is a very safe rule to rigidly reject all packings which require end-on compression to keep them steam-tight. The packing that ought to be preferred is the one that most nearly fulfils the following conditions, viz.: 1. It must keep the rod steam-tight with a minimum of friction. 2. It must be durable, but not liable to stick fast. 3. It must be easily removable. 4. It should not wear or scratch the rod. 5. It should be free to follow the rod, in case the neck bush should become slack, or the rod bent or out of line. 6. It should be capable of being adapted to old rods as well as new ones. 7. It should be easily adjustable, simple in construction, and automatic in its action. And 8. For remaining steam-tight it must depend upon some other principle than direct end-on compression.

CHAPTER V.

Lining and Adjustment of Bearings—Thrust Collars—Air-Pump—Circulating-Pump—Feed-Pumps—Condenser—Lubrication of Working Parts—Caution as to injurious Oils—Sight-feed Lubricator—Expansion Links—How to alter the HP by their Means—Feed-water Heaters—Explanation of their Economy—A Theoretically Perfect Feed-Heater—Weir's Feed-Heater and Evaporator—Morison's Evaporator and Feed-Heater—Board of Trade and Lloyds' Requirements.

ONE important element in the successful working of marine engines is to make sure that the crank-shaft bearings are truly in line with each other. The loss of efficiency and increased cost of working caused by bearings being out of line are often very serious, and in addition to this the risk of a broken shaft is very considerable. All bearings should be brass and brass, but not adjusted too finely; it is much better to have the shaft a little free than to require to slack back with a hot bearing. The practice of leaving the brasses open and screwing them up till sufficiently tight, and trusting to the set pins to keep the nuts from slackening back, cannot be too strongly condemned. On the other hand, the nuts should not be flogged with a big hammer until they will move no more, but ought to be reasonably tightened with a light hammer, so that they may be readily slackened back when required. It is a good plan to have the nuts marked as shown in Fig. 30; each cant

62 *Triple and Quadruple Expansion Engines*

of the nut is figured from left to right, and the top marked to correspond; these divisions can be again subdivided into any required number of parts, then by having a mark on the front of the bolt the position of the nut can be readily and correctly ascertained. The advantages of this will be readily recognised, especially on the score of neatness, as it does away with the barbarous and slovenly and eventually dangerous custom of hacking and scratching marks all over nuts and rings when taking leads or opening out for examination.

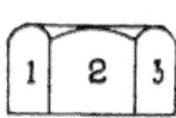

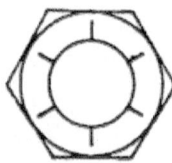

Fig. 30.

It is hardly necessary to notice all the working surfaces, but a few of the more important may be mentioned. Be sure that the crank and tunnel shafting are in line; if they are not, it is quite apparent that they will be bending every revolution, and putting an enormous strain on the coupling bolts. Besides the great risk of breaking down, there is a continual loss of power on account of the extra friction.

The thrust collars should be very carefully adjusted, so that each may take an equal share of the friction. The air-pump bucket should be a good working fit in the chamber. It is a bad plan to have the packing so tight that the bucket has to be hammered in, and trust to it working slack afterwards. It should be left slack enough to descend the chamber by its own weight. Never use tarry rope for packing the air-pump bucket; plaited flax rope yarn is perhaps the best packing that can be used in all ordinary cases.

Keep the air-pump valves in good order and free from grease, and if the bucket is fitted with metallic valves, see that they are properly bedded, otherwise there will be a loss of vacuum.

The circulating-pump piston or bucket should also be a good but easy working fit in the chamber. It requires no packing, but a good plan is to have grooves turned in it, which hold the water and thus form a sort of water packing. The circulating-pump ought to be fitted with air valves that can be adjusted so as to admit a sufficient quantity of air to lessen any shock that may be caused by the action of the water when the pump is at work.

The feed-pumps should be well looked after, and packed with good packing, as a leaky gland means a loss of pure water, which has, of course, to be made up with water containing solid matter, thereby increasing the liability to cause scale in the boiler. Never use canvas steam-packing for the feed or any other pumps having brass or Muntz metal rod or plungers, as it is too harsh and extremely likely to score them. There are a number of patent packings specially made for such rods, but, failing them, square plaited flax rope yarn is the best substitute, and is easily made. It is better where two feed-pumps are fitted to use one only at a time, say day about, or one on the outward voyage and the other on the homeward. One pump, if in proper order, is quite sufficient to maintain a full feed. By using only one ram the amount of wear and tear and of friction is reduced, and further, the quantity of compressed air pumped into the boiler at every stroke is decreased. This air, it must be remembered, is considered very injurious to the internal parts of the boiler.

Do not allow the lift of the feed-pump valves to increase too much, and do not open the feed-check valve on the boiler too wide, as that will only increase their wear and tear, and they will not work so steadily. Feed-pumps often give a great deal of trouble from too much valve lift, and when they are working irregularly it will generally be discovered that the suction valves rise too far, and the pumps perhaps, after two or three strokes, entirely clear the hot well of water, and for the next few strokes being choked with air will refuse duty, the over-flow being fully employed in the meantime. In ordinary circumstances it will be found that $\frac{1}{8}$ inch for the suction and $\frac{3}{16}$ inch for the delivery valve is quite sufficient lift. A pump with these lifts will deliver the water steadily, with very little clatter from the valves, and with very little wear and tear.

The condenser should be kept thoroughly clean. Allowing it to remain dirty simply means that more injection water will be required to condense the steam, more work will be placed upon the circulating-pump, and consequently there will be a loss of power to the engines. If caustic soda is used for cleaning the condenser tubes, they should be cleaned in port. When starting again the feed should be kept shut off until the condenser is entirely free from all dirty water. If common soda is used for the boilers, make a weak solution of it to prevent it taking any grease from the condenser into the boiler. Few things are more difficult to maintain or easier to lose than a good vacuum, it therefore behoves an engineer to remedy every defect that may be likely to produce an impaired vacuum. A few of the more important of these may be mentioned as follows: Dirty condenser tubes, defective air-pumps or valves, inefficient circulating-pump or valves, leaky LP' piston, slide valve, or glands, and leaky joints, cocks, valves, or pipes on LP cylinder, or on eduction pipe or condenser.

A considerable difference of opinion exists as to whether the pistons and valves ought to be lubricated, some maintaining that the moisture arising from condensation of the steam provides sufficient lubrication, and it is said that many engineers in charge of ordinary compound engines, carrying about 70 lbs. pressure, rely entirely upon sufficient steam condensing on the walls of the cylinders for lubricating purposes. This may be the case with such a pressure, but it has been found absolutely necessary in practice to provide additional lubrication to the cylinders when carrying the high pressures now in use; because the steam, being of so much higher a temperature, is very much drier, and consequently affords much less moisture for lubrication. Some engineers, no doubt, object to the use of internal lubricants on account of the injury they are likely to do to the boilers, and this may occur where the oil is of animal or vegetable origin, and is used to excess; but where a mineral oil is employed, and that sparingly, no evil results should follow. The cylinders, piston-rings, and valves should be examined every voyage for some time at least, and their condition will form the best guide to the proper quantity of oil to be used, keeping in mind that if more be admitted than is absolutely necessary the surplus merely fouls the condenser, or finds its way into the boiler. It is also of the utmost importance that only the very finest quality of mineral oil, such as that now supplied by several well-known makers, should be used, and there being so many worthless imitations of these high-class oils, resembling them both in colour and smell,

E

it is evident that too much care cannot be exercised in their selection. Even the best oil, as already mentioned, should be used very sparingly, as the internal moving parts require only a very small quantity of it when the engines are under weigh; but when going into or out of port there is usually more required, because when the engines are going slowly there is not enough steam used or condensed to keep the working parts sufficiently moist. Another reason for using oil sparingly is that if an excessive quantity has once been used it is extremely difficult to reduce the amount without doing injury to the cylinders, as many engineers well know who have tried it, the cylinder walls being almost certain to cut up should the quantity be suddenly reduced.

It is better in every case to have a sight-feed lubricator fitted to the cylinders, for by this means the engineer will be able to see that the oil is passing with due regularity into the cylinders, and to regulate to a nicety the quantity used. There are now quite a number of sight-feed lubricators which do their work very well, and it might be invidious to mention one in preference to another; but in every case the one that has the fewest working parts, and is the least likely to get out of order, should be selected, as nothing is more troublesome than to find when at sea that the lubricator will not work. The rods should be swabbed at least once an hour, and with the same kind of oil that is used for the cylinders, for some of it is sure to find its way inside, and that is a good enough reason why neither tallow nor any other kind of grease or oil should on any account be used for this purpose.

Where possible, automatic lubricators of a reliable

type should be used for the other working parts of the engine, as even with the most careful oiling by hand there is considerable waste. Although oil should be used in the most economical manner, the practice of running water on the bearings for the purpose of saving oil cannot be too strongly condemned. This may in some cases be due to shipowners refusing to send a sufficient supply, in others to engineers wishing to show how economical they can be with the stores; but, whatever the reason, it is a great mistake, for engines which have to overcome the enormously-increased friction arising from insufficient lubrication must necessarily give out considerably less power, while the loss from extra wear and tear of the working parts will greatly exceed any saving that might be effected in the amount of oil used. Besides this, many crank shafts have been utterly ruined by the indiscreet use of water, and, without considering the great risk of hot bearings, or anything else, the value of the oil that may be saved is but a trifle compared with the loss of speed following upon the increased friction. Suet is probably the best and most economical lubricant for tunnel bearings, as it lasts longer than tallow and does not run out of the bearings so readily. The plummer-blocks should, of course, be provided with save-alls, so that the suet which has run into them can be used over again, by being first melted and then strained into a vessel or mould, due care being taken that no grit is allowed to get into it.

The indicator is of the greatest assistance to the engineer, and diagrams should be taken at least every voyage, on the day before arriving in port. This will often save the engineer a great deal of unnecessary work and loss of time in overhauling pistons and slide valves, and this, perhaps, when the time could be ill spared from some other work of equal importance. It must be borne in mind, however, that the indicator does not show everything that is going on inside; there may be a flaw in the piston-rod or valve-spindle, or some other part about to fracture, of which the indicator does not give the least intimation. It behoves every engineer, therefore, to be constantly on the alert to guard against anything going wrong. The subject of indicator diagrams is a very large one, and is entirely beyond the scope of the present work, but it may be read up in any good text-book on indicator diagrams, and for our present purpose we must assume that the reader has at least an elementary knowledge of the subject. Before proceeding to take a diagram, see that the indicator pipes are perfectly clear, and the indicator itself thoroughly clean and in good order. Having done so, take two or three cards from each cylinder, as quickly as possible after each other, taking particular care that the steam, vacuum, link gear, etc., remain exactly the same. After the cards are obtained, proceed to work out the horse-power on each card, taking the mean of the two or three cards for each cylinder; this insures greater accuracy in the calculations. It is of the greatest importance that the three cylinders should develop exactly the same horse-power, or as near as it is possible to get them. The nearer this result is obtained, the more evenly will the engines run, the twisting moments upon the shaft will be more nearly equal, there will be a more equal amount of wear upon all the main bearings and top and bottom end brasses, and by this means the amount of overhauling required during the voyage will be considerably lessened. It has frequently been noticed that, when a certain top or bottom end was running badly, and required a good deal of attention, the corresponding engine was exerting more power than it was intended to exert, consequently the trouble arose inside the engine, and no amount of overhauling the bearing would remedy the defect.

The majority of engines are now fitted with separate expansion links, and a great variety of alterations can be made by an intelligent use of this means of linking up. It may be interesting to consider a few of the cases that often occur in which the cut-off can be altered without shifting the sheaves.

If the high-pressure valve-gear is linked up by itself the effect will be to decrease the total horse-power of the engines, but the same ratios, or very nearly so, will exist between the power in each cylinder, for, if we decrease the supply of steam to the HP cylinder, it will be decreased to the same extent in the others.

Were the intermediate gear alone linked up the result would be to decrease the power of the HP cylinder, and to increase that of the IMP cylinder, the LP remaining the same. This should be done when it is found that the HP cylinder is developing the most power, the LP next, and the IMP the least.

When the LP gear is linked up it increases the power of that engine, and decreases that of the IMP by causing it to have a greater back pressure, the HP remaining the same as before. This will, therefore, be a remedy when the IMP is exerting most power, the HP next, and the LP least. Another example is to link up both the IMP and the LP and keep the HP the same. This will increase the power in the LP, decrease the HP, while the IMP will remain as before, this being the remedy when the HP indicates the most power, the IMP next, and the LP least.

Where the links cannot be moved separately, a small amount of linking up can sometimes be effected by inserting liners between the ends of the drag links and the brasses. Linking up, if carried too far, is undoubtedly a bad practice, as it shortens the travel of the valve, and delays the opening and cut-off, thereby throttling the steam in its passage to and from the cylinder. It is, however, a very convenient way of altering the cut-off, provided a separate expansion valve is not fitted, or some other form of valve-gear which provides a variable cut-off.

If a considerable alteration of power is required, it will be necessary to alter the sheaves; putting the sheave forward and piecing the steam edge of the valve has practically the same effect as linking up. Should it be found that the LP is indicating the most power, the IMP next, and the HP least, the reverse of linking up must be done—that is, to carry the steam farther on the stroke. This can be accomplished by putting back the LP and IMP sheaves and chipping a piece off the steam edges of the valves.

Shipowners are now beginning to realise the commercial advantage of keeping their boilers free from scale, and so minimising the cleaning expenses, apart altogether from the fact that boilers so treated steam better and last much longer. And as feed-water heaters are now becoming more and more extensively used with triple expansion engines, it may not be out of place to point out wherein their economy consists, more especially as this question has so often been asked. To those who possess a correct knowledge of the principle of the compound engine this subject should present little or no difficulty, as the physical facts which govern the one have merely a further application in the other, and these have been briefly summarised by Mr. James Weir in his book on *Terrestrial Energy*, as follows:—

1. There is only one quality of steam. It may be superheated, but cannot be cooled below the temperature due to its pressure. Moist steam, unless the expression be taken to mean a mixture of water and steam, is an impossibility.

2. As much work can be got from exhaust steam as from steam taken direct from the boiler, provided the pressure is the same.

3. The value of steam is in direct proportion to its pressure.

4. Heat in any other form than steam in the engine is not only a loss, but is the exact measure of that loss.

5. The whole quantity of steam in the cylinder (as shown by the indicator) performs the theoretically possible amount of work, but this quantity does not represent the whole of the steam that enters the cylinder.

6. The difference between the quantity of steam entering the engine and that shown by the indicator is the amount of loss, *i.e.* the quantity of heat liberated by the *condensation of the steam* on the metal, which during exhaust is absorbed again by re-evaporation. The amount condensed is equal to the amount absorbed together with the work *transmitted* by the engine.

7. The proportion between the total quantity of steam

admitted and that shown on the card is dependent upon the difference of initial and exhaust temperatures and the weight of steam used per minute taken in relation to the surface exposed.

8. Each cylinder of a compound engine is a simple engine.

9. The efficiency of every engine depends on getting the initial pressure on the piston as near to the boiler pressure as possible, and returning the feed-water as near the exhaust temperature as possible.

10. In all condensing engines the feed-heater is the condenser. Every steam engine, to perform or *transmit* work, must receive steam at a pressure greater than that at which it exhausts, and it must always exhaust steam, because each simple engine can only use a *very small proportion* of the *work* in the steam, no matter how great the pressure or expansion may be.

11. There are only two ways of getting heat into a steam boiler—the one by conduction through the metal, the other by the feed-water. That by conduction must be at a higher temperature than the steam, but that by the feed-water may be used at a lower temperature, hence some of the exhaust waste may be put into the boiler by the feed-water.

12. All the heat put into the boilers from the fuel (radiation and other preventable waste excepted), less the heat transmitted by the engine in the form of work, must be exhausted as steam.

13. The heat that is carried into the boiler by the feed represents an excess that has come out of the boiler and passed through the engine. There is, however, none of this wasted, as it is all returned to the boiler, repre-

senting a quantity of steam wrought in a theoretically perfect engine.

14. Exhaust steam, or waste heat, must be used to effect economy. By using steam direct, the result will be neither loss nor gain, but the effective amount of steam the boiler will produce without priming will be diminished in proportion to the quantity taken from it to the feed.

There have been numerous methods tried, with more or less success, for heating the feed-water, some by endeavouring to utilise the waste heat in the funnel, others by means of steam direct from the boiler, and Weir's system, viz. by using a portion of the steam in the low-pressure receiver. The latter method is generally considered to be the most efficient, and is an undoubted economy, according to Prof. Cotterill, who in his note on feed-water heaters says:—

It is the purpose of this note to show that a feed-water heater may be so designed as to play the part of a regenerator, and to point out that this is the cause of the saving due to certain feed-water heaters employed in practice.

For suppose the feed-water as drawn from the hot well by an ordinary feed-pump to pass into the boiler by a supply-pipe in which a number of gratings are placed as in the regenerator of an air engine. These gratings, however, are now to be supposed tubular, and each of them is connected by its own pipe with the cylinder, so that steam may be admitted at pleasure by a suitable valve. This valve will be supposed so constructed that steam enters only at one particular pressure, and the pressures in the several gratings are so adjusted as to increase gradually up the pressure in the boiler. To each pressure corresponds a certain temperature, and therefore the temperature of the gratings will gradually increase from the temperature of the hot well to the temperature of the boiler. Each grating has, moreover, a small orifice at its lowest point, so as to allow the drainage of the gratings due to the steam condensed in them to escape and mingle with the original feed-water. The effect of these arrangements will evidently be that the feed-water passing through the supply-pipe will have its temperature raised as it passes through each grating in succession, and together with the drainage of the gratings will ultimately enter the boiler at the temperature of the boiler. The heat expended in the boiler per pound of steam will now be the latent heat of evaporation only.

A feed-water heater thus constructed would be theoretically perfect, and the engine which possessed it would, if otherwise perfect, have an efficiency given by the simple formula which applies directly to a Carnot cycle. For such an engine receives no heat from the source of heat except at the temperature of the boiler, while every step of the process is strictly reversible. The tubular gratings serve exactly the same purpose as the regenerator of a Stirling air engine, the only difference being that the processes of storing and re-storing heat go on continuously instead of taking place alternately. The heat supplied by each grating to the feed-water is re-stored at exactly the same temperature by steam which condenses there. Thus we learn that the "loss by misapplication of heat to the feed-water" in a steam engine can be avoided not only by the ideal pumps described in theoretical treatises which have never been used in practice, but also by a properly designed feed-water heater, which increases the efficiency of the engine exactly as a regenerator increases the efficiency of a Stirling air engine. Instead of the gratings we may suppose the supply-pipe jacketed in divisions, each division being supplied with steam and drained as before.

Let us now consider the feed-water heater introduced by Mr. Weir, which is supplied with steam from the low-pressure receiver of a triple expansion engine. If the steam, instead of being taken from the receiver, were taken from the boiler, the feed-water might be raised to the temperature of the boiler, but the loss of work which might have been done by the condensed steam would exactly compensate for the saving of heat, so that the process on the whole would be neither a gain nor a loss. If, on the other hand, the steam were taken at release from the low-pressure cylinder, the feed would only be raised to the temperature of release, but the saving of heat would be an unmixed gain. It is therefore easily understood that if the steam be taken from the low-pressure receiver, that is, after it has done two-thirds of its work, there must be on the whole a gain. For example, suppose 10 per cent. of the whole expenditure of heat is saved by raising the temperature of the whole amount of feed-water up to the temperature of the receiver. Making an allowance for the amount of suspended moisture which the steam will contain, it may be estimated that about 12 per cent. of the whole amount of steam used will be drawn from the receiver. The work which would have been done by this steam in the low-pressure cylinder may be taken as one-third of the whole, that is, as 4 per cent. of the whole power of the engine. Thus the efficiency is increased about 7 per cent.

Though this reasoning seems clear, yet so long as the nature of the saving thus obtained is not understood, doubts may be felt whether the saving is a real one, especially as the exact amount cannot be calculated in the absence of experiment. It might be supposed that there is a compensation in diminished action of the steam which is not used in the feed-water heater. On comparison, however, with the theoretically perfect heater previously described, the nature of the saving becomes clear, for the process is evidently equivalent to the employment of a regenerator with a single grating, by means of which the efficiency of the cycle is increased, approaching more closely to that of a theoretically perfect heat engine. Of course the whole gain due to a theoretically perfect heater will not be realised, but a considerable fraction, probably not less than one-half, may be obtained. If the engine be supposed in other respects perfect, an exact calculation may be made of the gain: in an actual engine it would be necessary to know the weight of steam used by the feed-water heater, and even then exact comparison is difficult, because the subtraction of steam from the low-pressure receiver virtually alters the proportions of the cylinders. With the high pressures used in recent practice it may reach 12 per cent. or even more.

Having mentioned this feed-heater, and knowing there are many engineers who are not acquainted with it, we felt justified in describing the firm's latest designs for large triple expansion engines, as fitted to the Hamburg-American liner *Normannia*, consisting of feed-heater, feed-pumps, and evaporator, as shown in Figs. 31, 32, and 33. The feed-heater proper consists of an upright cast-iron cylinder, with two compartments, an upper and a lower. On the cover there is a spring-loaded valvet for admitting the cold feed-water, and a large non-return valve on the side for admitting the steam taken from the intermediate cylinder exhaust. The main engine feed-pumps are used to take the feed-water from the hot well to the heater. It passes through the spring-valve on the cover in a thin sheet and gets instantly heated by contact with the steam, and by this heating the air in solution in the water is liberated and driven up to the air vessel on the top, whence it is removed to the condenser or atmosphere through the small cock on the side of the air vessel. The water in the bottom of the heater is then at the boiling temperature due to the pressure in the heater, and is free from air. In this state it has no corrosive effect on iron or steel, and to get it into the boilers in this condition a float is placed in the lower part, which controls the speed by regulating the supply of steam to drive the feed-pumps; the level of the water is thereby kept constant in the heater, and the pumps are completely filled with water.

The feeding engines are of the direct-acting type, and one of their special features is the slide valve of the steam cylinder. This is an ordinary three-ported valve kept up to the face by the steam pressure. On the back of this valve are expansion ports and also ports for admitting steam to the ends of the slide valve, which works into hollow cylinders slipped on the round ends of the main slide and held in place by the covers. On this flat face the auxiliary slide, which is an ordinary D valve, works. This distributes the steam to each end of the main slide alternately, and the outer edge cuts off

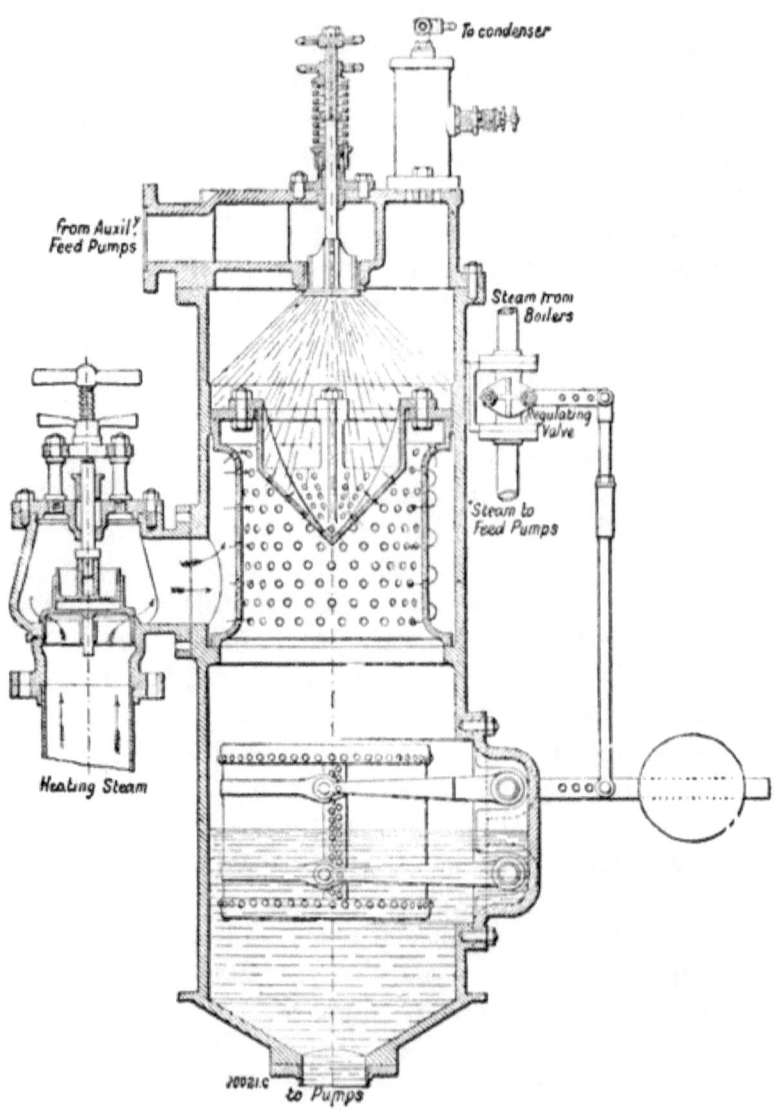

Fig. 31.

the steam entering the expansion ports at 8 inches from the end of the stroke. The auxiliary valve is worked by levers driven by the piston-rod.

The operation is as follows: When the piston is at the end of the stroke, the auxiliary slide opens the exhaust to one end of the main slide; the other end being in this position open to the steam, the valve is

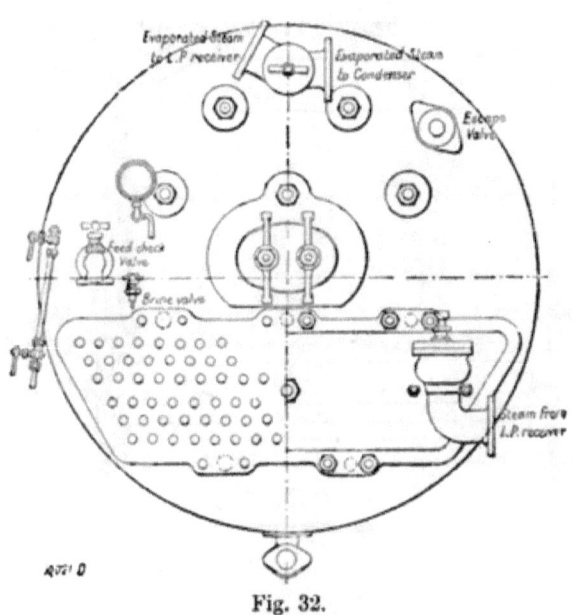

Fig. 32.

thrown over until the exhaust is cut off, which acts as a cushion. In this position the main and expansion ports are full open for the return of the piston, which moves at a rapid rate until the expansion is closed; then at 8 inches from the end the auxiliary valve closes the expansion ports, and the speed is reduced towards the end of the stroke, when the auxiliary valve opens the exhaust and throws the main slide for the return stroke. The expansion chambers are fitted with by-passes to admit the steam full stroke. This is necessary when starting, as then the cylinders are full of water.

The steam pistons are made of steel and the water pistons of gun-metal, the piston-rods being of cold-rolled manganese bronze.

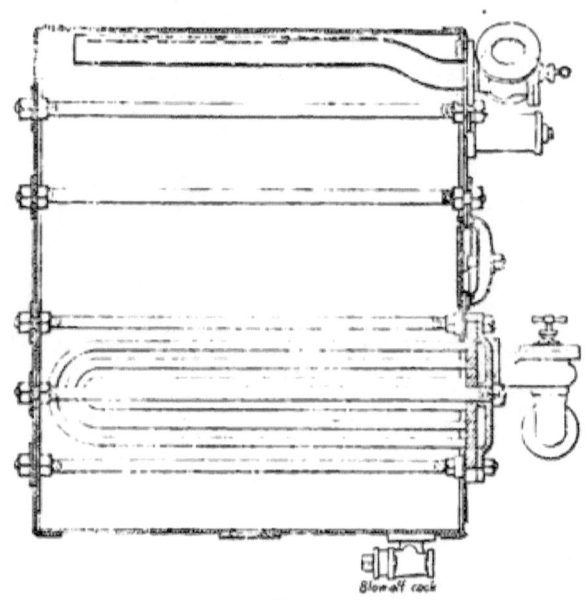

Fig. 33.

Another special feature is the valve arrangement in the water end. The valves are of the patent group type, capable of giving a large opening with a small lift for the free flow of the hot water. There are twelve in each suction and seven in the discharge seat, thirty-eight valves in all. Each pump is perfectly independent of the other, and they are arranged to discharge to the boilers, either through the main or the auxiliary feed pipes, so that an accident to one line of pipe cannot interfere with the regular supply of water to the boilers.

There are two evaporators capable of evaporating 80 tons of sea-water per day, for supplying the main boiler with fresh water. Their special features may be shortly stated thus: They are triple effect in action, as the steam admitted to the heating tubes is taken from the exhaust of the high-pressure cylinder, and the steam raised from the sea-water is led into the low-pressure receiver, the loss of heat for the production of the fresh water being confined only to the intermediate cylinder. Therefore, two-thirds of the work is saved, which would be lost were the steam taken from the boiler direct and the evaporated steam put into the condenser.

A point calling for special mention is the arrangement for making the tube service uniformly efficient—a desideratum where a number of tubes are used. The tubes are U shaped, the ends in the admission chamber being open, while the outlet ends are closed by plugs or plates having only a small hole as an opening to the outlet chamber. The water of condensation is again led by tubes through the evaporator before it escapes into the hot well. By this means there is a lower pressure at the outlet ends, and a constant current is kept up through all the tubes, preventing any accumulation of air or water.

Another feature to be noticed is that the exhaust steam from the feed-pumps and the steam from the evaporator is led into the pipe connecting the heater with the low-pressure receiver, and where there is no auxiliary condenser the exhaust from all the auxiliary engines on board is led there; thus making all the simple engines compound and the compound engines triple expansion. By this simple means, considerable economy is effected without any extra complication. When the feed-heater is working the steam goes to the heater, and when shut off, coming into or leaving port, it goes to the receiver. Circulating-pumps, engines, driving fans, refrigerators, or dynamos, work steadier, and with less wear and tear, than when exhausting direct into the condenser.

Another very efficient evaporator is that known as Rayner's, which may be described as follows:—

It consists of a vertical cylindrical vessel, in the lower part of which is fitted a series of horizontal coils of solid drawn copper tube, forming the heating surface. These coils are attached to suitable passages formed in the door of the evaporator, at their outer ends, and each pair are similarly connected at their inner ends, the passages in the door and the central connecting pieces being so arranged that steam admitted to the top coil has to traverse that one, and then passes down to the second, and from second to third, and so on, till it has traversed the entire length of all the coils in succession.

The steam passing through these coils parts with its heat to the sea-water (which is admitted to the lower part of the evaporator), and causes it to boil and give off steam, which, rising to the upper part of the evaporator, is led by suitable valves and pipes to the condenser of the main engine, where it is condensed, forming auxiliary feed-water, which is pumped to the boilers by the ordinary feed-pumps.

The illustrations that follow give a very clear view of the form of the evaporator, and of the heating coils and their connections.

The special form of the heating coils, which are horizontal helices, secures their self-cleaning property, and this maintains their efficiency.

Fig. 34.

These coils being the hottest part of the apparatus, the mineral deposits thrown down by the boiling sea-water tend to adhere to them; but from the constantly varying temperature outside and inside the coils, they expand and contract after the manner of a Bourdon gauge tube, and this action cracks off the deposited scale, which falls to the bottom of the evaporator, and is removed from time to time by the hand holes provided for that purpose. The door to which the coils are attached seldom requires to be removed; but should it be desired at any time to examine the interior, only the joint of the door itself has to be broken, and special handling gear is provided on all but the smaller sizes, to facilitate the handling of the door with coils attached.

It is most important that the level of the water to be evaporated should be maintained at a constant level, and that there should be no possibility of the water accumulating in the evaporator, so as to fill it, or pass over into the condenser.

This requirement is fully secured by our automatic feed inlet valve, which is found perfectly reliable.

This valve is contained in the chest seen on the side of the evaporator in the illustrations.

Owing to the large steam room provided, and a special arrangement of baffle plates in the upper part of the shell, no priming takes place in these evaporations.

The space occupied by these evaporators is very small, and only space enough to permit the withdrawal of the coils is necessary in front.

Radiation of heat to the engine-room is effectually prevented by the evaporators being well covered with hair felt, over which is fitted planished sheet steel, secured with brass bands. This gives the apparatus a highly-finished appearance, and keeps the engine-room cool.

The advantages of using the automatic evaporator may be summed up, as follows:—

Evaporation takes place under a vacuum of, say, 15″

at about 170°, whereas at 15 lbs. pressure per square inch it takes 250° to effect same result.

The water supply being taken from the circulating water, leaving the main condenser usually at about 100°, its temperature has only to be raised some 70° in the evaporator to produce vapour. Only this amount of heat has to be given off by the high-pressure steam in the coils, the remaining heat being communicated to the feed-water by the drain from the coils mingling with it.

The vapour generated in the evaporator gives off its heat to the condensed water in the main condenser, thus raising the temperature of the feed-water.

Another important advantage in evaporating sea-water under vacuum, at a temperature not exceeding 170°, is that the scale is not so hard as at higher temperatures under a pressure.

No special pump is required to feed this evaporator if fitted as we recommend, well down in the engine-room, below the level of the circulating-pump discharge pipe. The evaporator is worked under a vacuum which draws in the feed-water as required.

The evaporator may be worked whilst the main engines are at rest, by circulating a small quantity of water through the condenser. The water produced being run into storage tanks or pumped into the boilers.

The automatic evaporator requires less fitting up than any other other; the connections are all comparatively small. It is usually placed on the engine-room floor and needs no special fixing, bolting to the floor or a stool made up for it being all that is necessary.

The following instructions for working the evaporator are given:—

1. Open valve on circulating-pump discharge pipe,

admitting water to the automatic valve chest, water will flow into the evaporator until the proper level is reached, when the automatic valve will close.

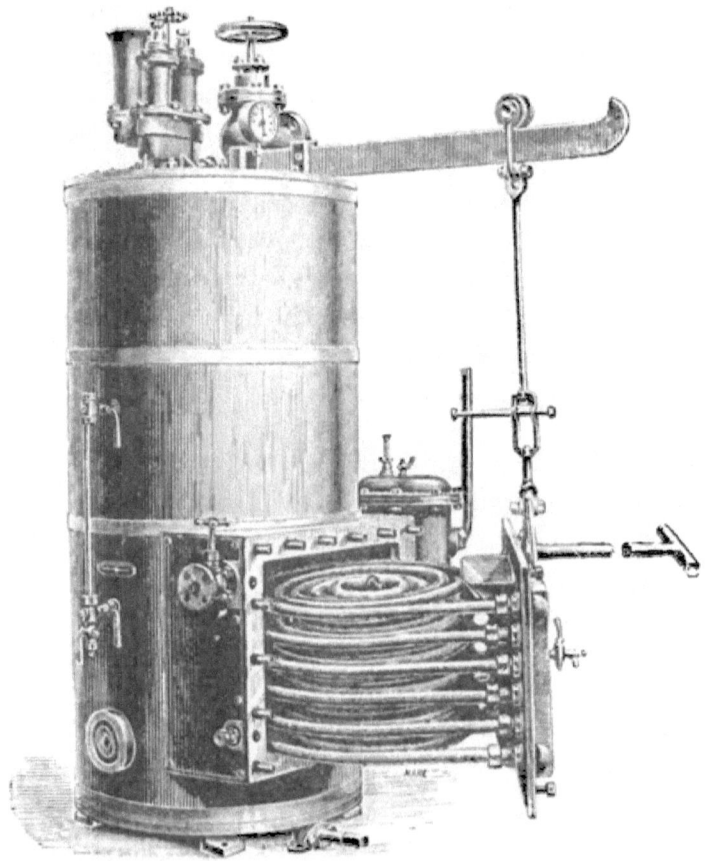

Fig. 35.

2. When the water appears in the gauge glass, open the vapour valve, putting the evaporator in connection with the condenser.

3. Open drain-cock from coils.
4. Open steam stop valve gradually.

A vacuum of from 5" to 15" should show on gauge, this can be regulated by the vapour valve.

At first the engineer should watch carefully at what vacuum he gets the best results, and regulate the opening of the vapour valve and steam valve accordingly. The various pressures of steam in use on different steamers necessitate this adjustment.

It may not be found necessary to open the steam inlet valve to its full extent. We have found instances (especially when the coils were perfectly clean) when, with the steam valve full open, the ebullition of the water was too fierce, and there was a tendency to prime. This was entirely cured by partly closing the steam valve.

The evaporator being at work will continue so, and requires no further attention except to blow off the brine periodically. The density of the brine should never be allowed to exceed 4/32, as ascertained by salinometer. The test-cock on the door is provided for this purpose.

To blow off the brine, shut vapour valve (thus destroying the vacuum) and open blow-off cock; the pressure which will quickly arise in the evaporator will blow out all the brine; any excess of pressure being relieved by the safety valve. When all the brine is discharged, shut blow-off cock and open vapour valve.

Do not shut steam off the coils when blowing off, it is unnecessary; but shut drain from coils and keep it closed till the water shows again at working level, then open it.

After blowing off, if it is desired to admit the water more quickly, press down the rod projecting through the top of the automatic valve box, and keep it down till the water shows proper level in the glass (as marked on the index plate), then pull up the rod as far as possible.

This rod is also useful in ascertaining at any time whether the float on the automatic valve is acting properly or not.

The scale which is thrown off by the coils falls to the bottom of the evaporator. This should be removed every few days by the hand holes provided for that purpose. If this is done, and the working instructions are attended to, the door, with the coils attached, need very seldom be removed at all.

Besides the foregoing, it is only fair to state that Morison's patent evaporator and feed-heater are also well and favourably known, and as these have already been fitted to several hundreds of steamships, a description of them may be interesting to our readers.

In designing the apparatus, there is no doubt every care has been taken to meet the requirements of actual work at sea, special attention having been given to simplicity, accessibility, and strength; and if any difficulty is experienced, resulting, perhaps, from an error in fitting in place, it should always be remembered that the success of the apparatus has already been established, and whatever the difficulty may be, it should be at once reported to the makers and remedied.

Regarding the amount of water necessary for auxiliary feed, it has been shown by very careful experiment that the actual loss in triple expansion engines varies from 4 to 8 per cent. of the total amount of steam used. Consequently, in an engine using 15 lbs. of water per IHP per hour, the total loss may be from $6\tfrac{1}{2}$ to 13 tons per day for each 1000 IHP developed. A safe allowance being 1 ton per day for each 100 IHP, and if evaporators are fitted of this capacity the results will be satisfactory under all conditions.

An evaporator may be arranged in various ways, depending upon the manner in which the vapour generated therein is to be utilised. The three principal methods of using the vapour are :—

a. By discharging it direct into the condenser, there to mix with the condensed steam, and be pumped into the boilers comparatively cold.

b. By discharging it into the low-pressure valve casing, so that it does work upon the low-pressure piston before being finally condensed with the main body of steam passing through the engine.

c. By utilising it to heat the main feed-water by means of a feed-water heater. In the case of Morison's patent the water is delivered into the boilers at a temperature of 150° to 170° Fahr.

The comparative economy of these different arrangements is given below, and it will be seen that the most economical method of working an evaporator is to utilise the generated steam to heat the boiler feed-water. The cost of evaporating a ton of water (neglecting loss by radiation, which is practically the same in each case) may be calculated as follows :—

a. Evaporating into the Condenser.

As it is usual to place a vapour or reducing valve between the evaporator and the condenser, it may be assumed that evaporation takes place at atmospheric pressure; and the temperature corresponding to this is 212° Fahr.

Triple and Quadruple Expansion Engines

The water for feeding the evaporator being taken from the circulating discharge, its temperature may be taken at 80° Fahr.

To maintain the density at $2\frac{1}{2}/32$, two-fifths of the total amount of water admitted to the evaporator must be discharged into the bilge, and consequently to produce 1 lb. of pure steam $1\frac{2}{3}$ lbs. of water must be admitted to the evaporator, of this 1 lb. is evaporated and $\frac{2}{3}$ lb. are discharged into the bilge as hot brine.

The total heat required to make 1 lb. of pure steam is the sum of the heat in the steam and that discharged in the hot brine.

Amount of heat, in thermal units, in 1 lb. of steam at 212° above 80°, ... } 1099 thermal units.

Amount of heat in $\frac{2}{3}$ lb. brine at 212° above 80°, . . . } $\frac{2}{3}$ of $132 = 88$,,

Total heat required to make 1 lb. of pure steam, being the sum of the above, ... } $1099 + 88 = 1187$,,

If it be assumed that 1 lb. of coal will evaporate 10 lbs. of water from and at 212° Fahr., that is, that 9666 thermal units will be obtained from the combustion of 1 lb. of coal, it follows that the amount of pure steam generated by 1 lb. of coal with this arrangement is—

$$\frac{9666}{1187} = 8.14 \text{ lbs.}$$

Consequently, to make 1 ton of fresh water the amount of coal required is—

$$\frac{2240}{8.14} = 275 \text{ lbs.}$$

b. EVAPORATING INTO LP VALVE CASING.

In this case the steam from the evaporator is discharged into the low-pressure valve casing, and does work upon the low-pressure piston before being condensed. There is therefore a saving over the previous arrangement, which may be calculated as follows:—

In an economical triple expansion engine 17 per cent. of the total heat in the steam is turned into work, consequently if it is assumed that the power developed in each cylinder is equal, one-third of this (or 5·66 per cent.) is developed in the low-pressure cylinder.

Also, as 17 per cent. of the total heat is utilised, 83 per cent. must be rejected; the amount of heat entering the LP cylinder must therefore be 88·66 per cent. of the total heat.

That is to say, out of the total heat entering the LP cylinder the amount utilised is—

$$\frac{5\cdot 66 \times 100}{88\cdot 66} = 6\cdot 38 \text{ per cent.}$$

Had this heat been used in the same manner as the rest of the heat passing through the engine, the amount utilised would have been 17 per cent. The amount regained in the LP cylinder is therefore—

$$\frac{6\cdot 38 \times 10}{17} = 37\cdot 5 \text{ per cent. or } \tfrac{3}{8}.$$

Now, let the pressure in the LP valve casing be 7 lbs. per square inch, the temperature corresponding to which is 233·1° Fahrenheit, and let the

92 *Triple and Quadruple Expansion Engines*

temperature of evaporator feed-water be 80° as above, then

Amount of heat in 1 lb. of steam at 233·1° above 80°,	1104·5	thermal units.
Amount of heat in ⅔ lb. of brine at 233·1° above 80°,	⅔ of 153·1 = 102·1	,,
Total heat required to make 1 lb. of pure steam, being the sum of above,	1104·5 + 102·1 = 1206·6	,,
Amount regained in L.P. cylinder,	⅜ of 1104·5 = 414·2	,,
Net cost of making 1 lb. of pure steam is,	1206·6 − 414·2 = 792·4	,,
Amount of steam generated per lb. of coal,	$\frac{9666}{792\cdot4} = 12\cdot19$ lbs.	

Consequently, to make 1 ton of fresh water the amount of coal required with this arrangement is—

$$\frac{2240}{12\cdot 19} = 183\cdot 7 \text{ lbs.}$$

c. Evaporating into Morison's Patent Feed-water Heater.

In this case the steam pressure in the evaporator is about 1 lb. above the atmosphere, and as the steam is entirely condensed among the feed-water, and is thus pumped into the boilers, the only actual expenditure of heat is that caused by brining.

Let the steam pressure in the evaporator be 1 lb., the temperature corresponding to which is 216·3° Fahr., and let the temperature of evaporator feed-water be 80° as before, then

Heat in ⅔ lb. of brine at 216·3° above 80°,	⅔ of 136·3 = 90·0 thermal units.

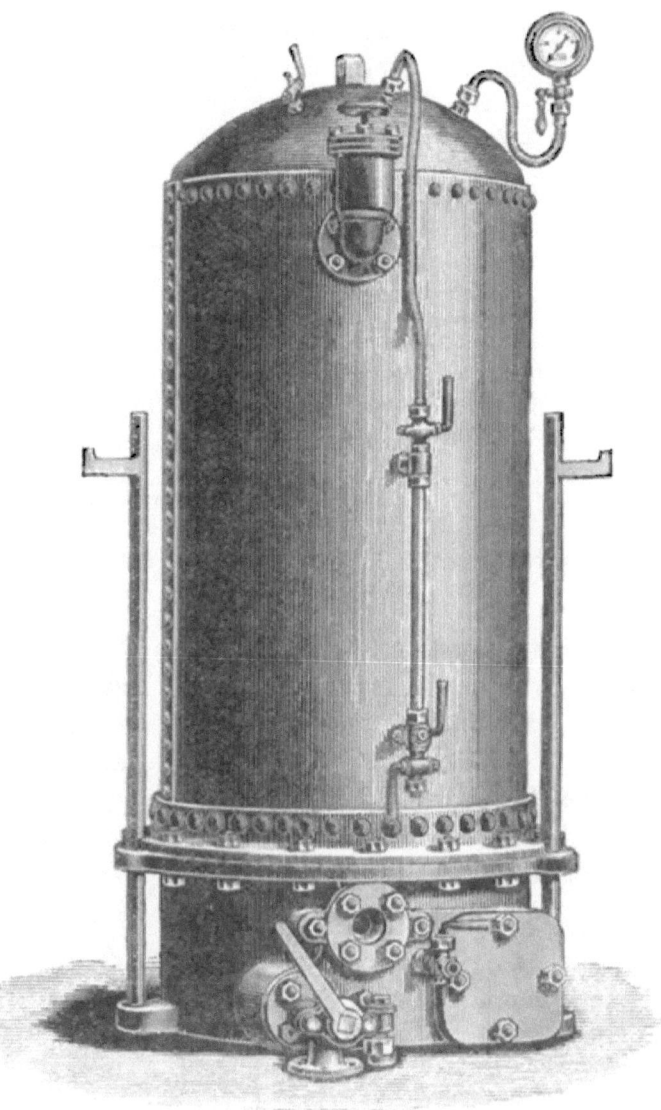

Fig. 36.

This being the whole expenditure of heat, the amount of steam generated per pound of coal is—

$$\frac{9666}{90\cdot 9} = 106\cdot 3 \text{ lbs.}$$

Therefore, the amount of coal required to make 1 ton of pure water is—

$$\frac{2240}{106\cdot 3} = 21 \text{ lbs.}$$

TABLE OF COMPARATIVE RESULTS, SHOWING THE AMOUNT OF COAL REQUIRED TO MAKE ONE TON OF PURE AUXILIARY FEED-WATER.

Method of Evaporation.	Pounds of Coal required to produce 1 Ton of Pure Water.
Into condenser,	275 lbs.
,, LP valve casing,	184 ,,
,, Morison's patent feed-water heater,	21 ,,

The following is a description of this evaporator:—

Fig. 36 shows the apparatus in working position.

Figs. 37 and 38 are sectional elevation and plan respectively. By breaking the joint of the dome to the base, the dome can be lifted clear of the coils and held in that position by means of the three guides or supports.

The various details are as follows:—

    A The dome of Siemens Martin steel, with stamped steel ring forming the flange which connects the dome to the base.

    B The safety valve.

    C The pressure-gauge.

    D The water-gauge, to show the height of the salt water within the evaporator.

    E An eye-bolt for lifting the dome.

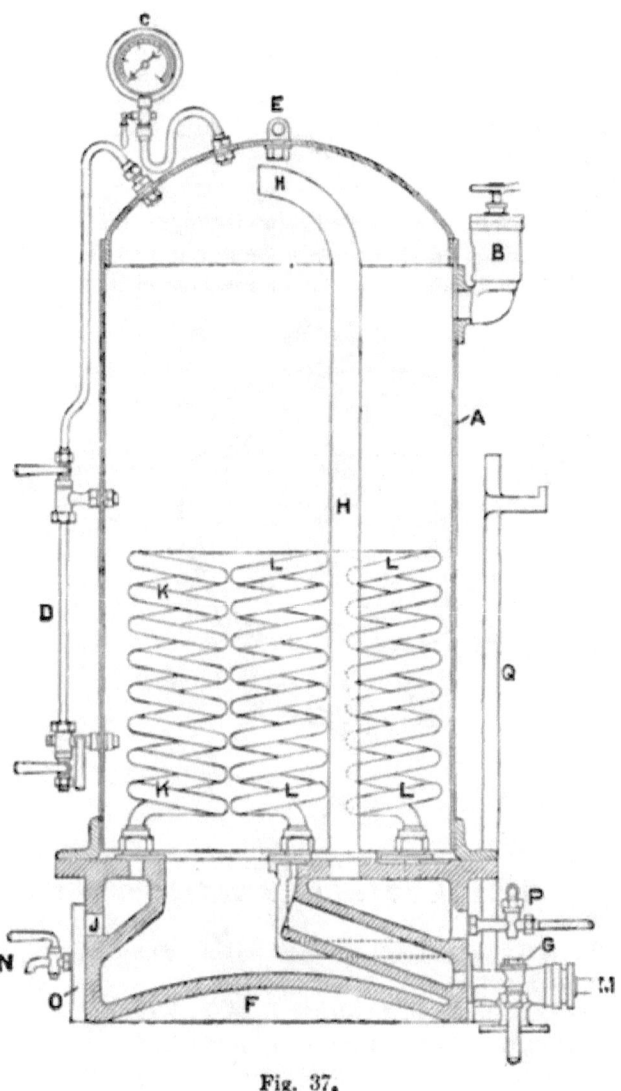

Fig. 37.

It should be noted that none of these fittings are disturbed when the dome is lifted.

    F The base or salt depositing chamber.
    G Brass non-return valve for salt-water inlet.
    H Copper internal pipe for outlet of steam generated in the evaporator.
    J Steam inlet.
    K, L The heating coils of solid drawn copper tube, secured to the lower vessel by special brass unions, which enable the coils to be removed quickly and conveniently.

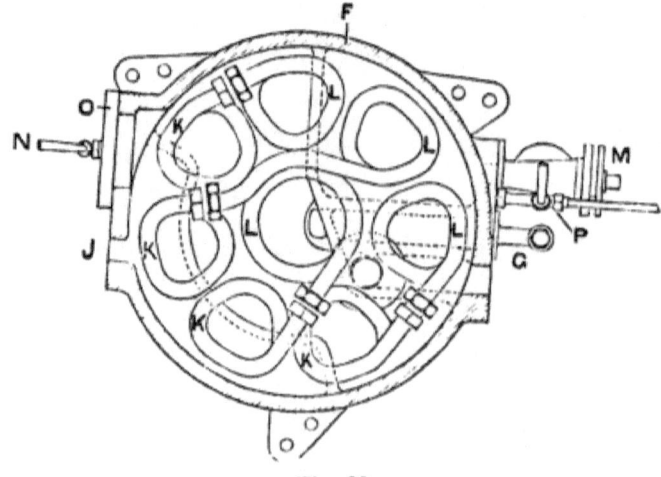

Fig. 38.

    M Brass cock for discharging brine.
    N Brass salinometer cock for testing the density of the water.
    O Door for cleaning out salt or other deposit, fitted with Muntz metal studs and brass nuts.
    P Brass cock to regulate drain from coils. (*Note.*—As the hole in the plug of this cock is just large enough to pass the water condensed in the coils, care should be taken that it is kept free from dirt.)
    Q Malleable iron guides and supports for dome.

As to the working of the evaporator it may be mentioned that it is exactly the same in all arrangements, and is as follows, viz. :—

Sea-water is admitted into the evaporator until the required level is indicated in the gauge glass.

Steam is then turned on at J, and in a few minutes the water in the evaporator will boil.

The density of the water in the evaporator requires attention, and should not exceed $2\frac{1}{2}/32$ (at the same time the density must not be kept much below $2\frac{1}{2}/32$, or the scale formed in the coils will be very hard).

The brine cock M should be regulated to maintain a steady discharge into the bilge, and in addition the whole of the water in the evaporator should be blown out at least once every four hours.

Very little experience will show how far the small valve on the brine cock should be open to maintain the required density, and when once adjusted, no further attention will be necessary.

When blowing out, steam should be turned off until the water has been all blown out, then turned on again suddenly; this cracks off the scale from the coils.

The cleaning of the coils presents no difficulty, and takes but little time, even after forty days' continuous use.

Fig. 39 gives a view of Morison's patent feed-water heater, which consists of an outer cast-iron vessel into which the steam from the evaporator is admitted, and, becoming condensed among the feed-water, imparts its heat to the latter. Its action is as follows, viz. :—

98 *Triple and Quadruple Expansion Engines*

The water from the hot well enters the heater by CW and the steam from the evaporator by SM.

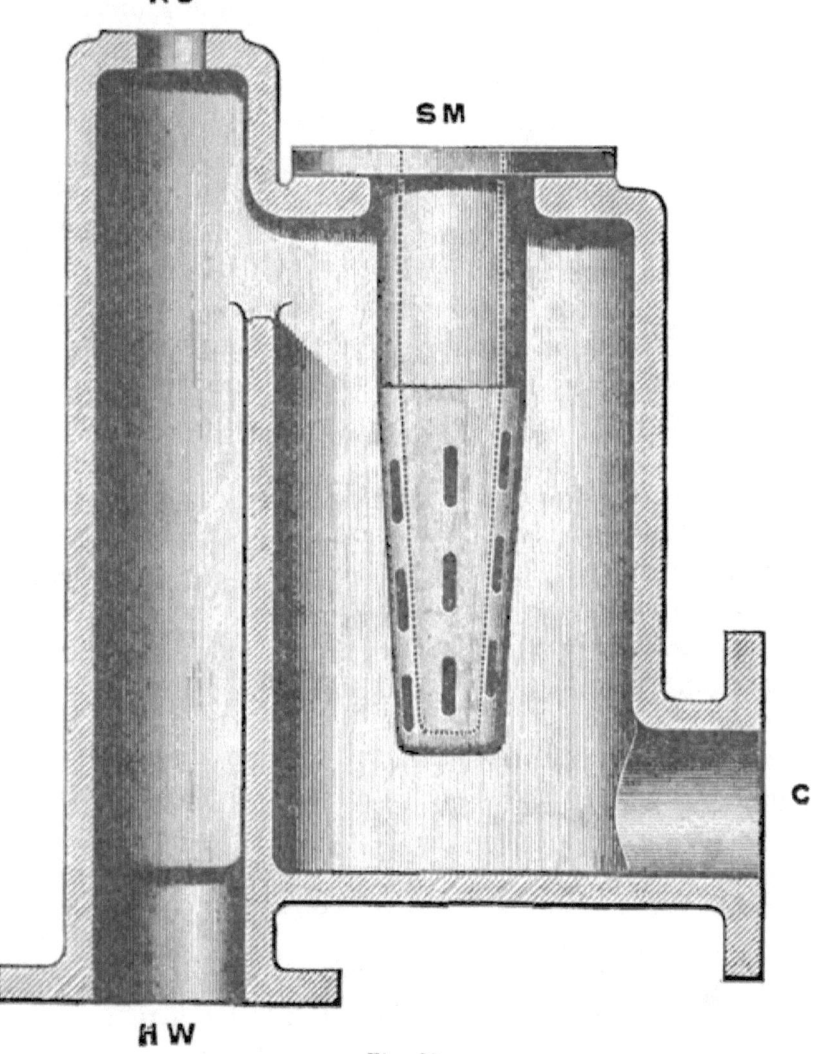

Fig. 39.

The slots in the internal pipe SM being set at an angle, the escape of the steam causes a centrifugal motion of the water among which it is condensed.

The heated water flows out at HW to the feed-pump suction, any steam or vapour generated within the heater being allowed to escape by AO to the hot well at any point above the water level.

The feed-heater should be so placed as to divide the distance from the air-pump to the feed-pump suction into two equal parts, so that the vertical distance from the air-pump head valve to the feed-heater will be about equal to the vertical distance from the feed-heater to the feed-pump suction.

In some cases, where the main feed-pump valves are exceptionally heavy, there may be difficulty in getting the pumps to work when using very hot feed-water. An almost certain remedy for this is to put a small cock on the main feed-pump valve chests, between the suction and delivery valves, and immediately under the discharge valves, and to lead a $\frac{3}{8}$-inch pipe to the condenser. This will clear the pumps of the vapour which impedes their action.

When evaporating into the condenser it is necessary to provide a reducing valve, so as to maintain a steady pressure within the evaporator, and thus avoid priming. This valve is shown in Fig. 40, and when in use the steam from the evaporator enters at B2, lifts the double opening valve D2, against the pressure of the spring E2, and escapes to the condenser by C2. The spring may be adjusted to any required pressure by the set screw F2, and is usually set to maintain a slight pressure in the evaporator.

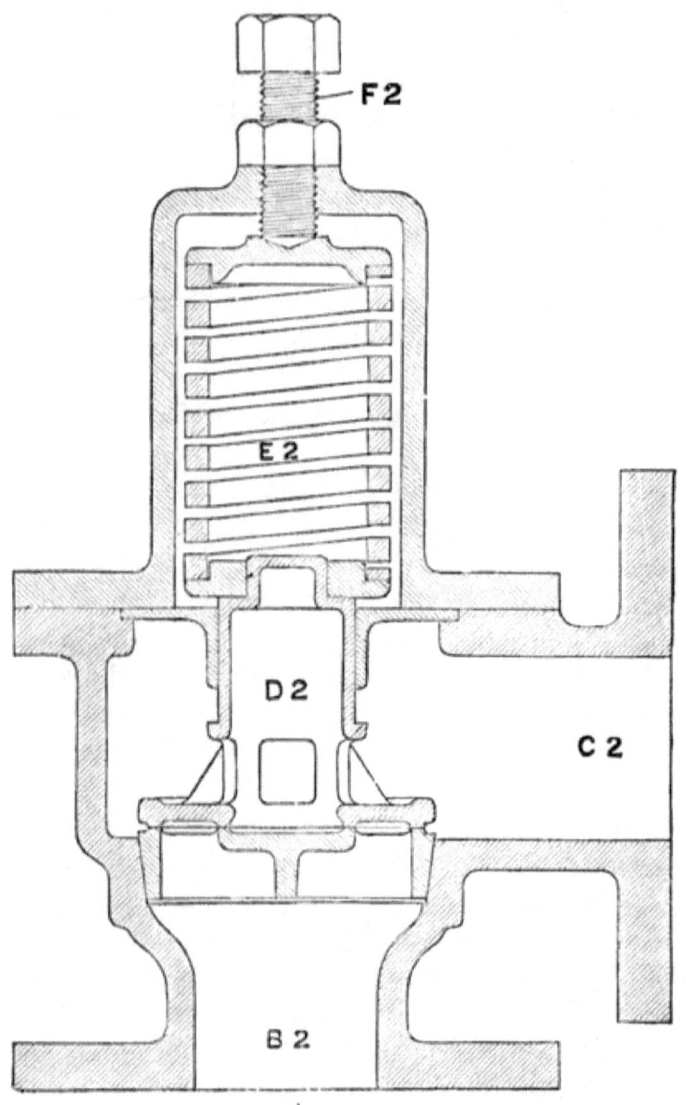

Fig. 40.

A brining cock is also arranged for continuous brining into the bilge, and in order that it may be readily adjusted to the amount required to maintain a density of $2\frac{1}{2}/32$ in the evaporator, a small by-pass valve is provided, as shown in Figs. 41 and 42. In ordinary working the plug is completely shut, and the small valve regulated to pass the necessary quantity of water. When this valve is once adjusted it need not again be touched, the

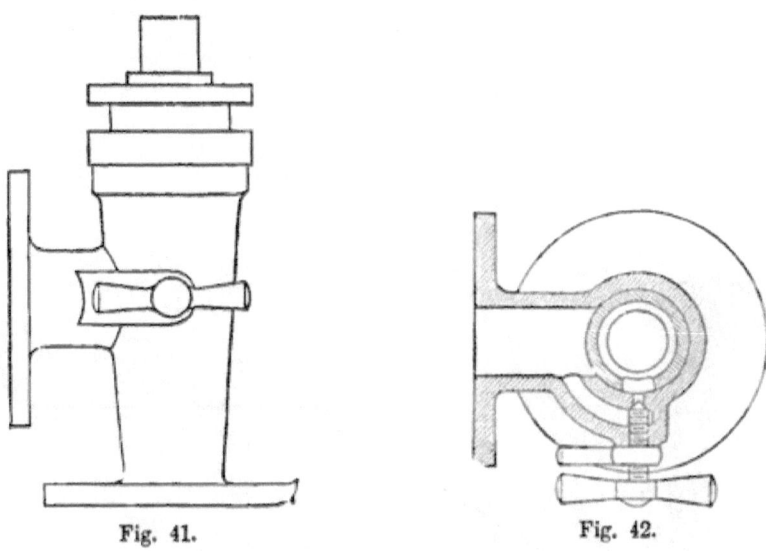

Fig. 41. Fig. 42.

plug being used for blowing the whole of the water out as previously described.

The following is the method of working the evaporators:—

a. EVAPORATING INTO CONDENSER.

Fig. 43 shows the simplest arrangement for fitting an evaporator; it is, however, the least economical.

102 *Triple and Quadruple Expansion Engines*

In this arrangement the steam generated in the evaporator is discharged into main condenser through

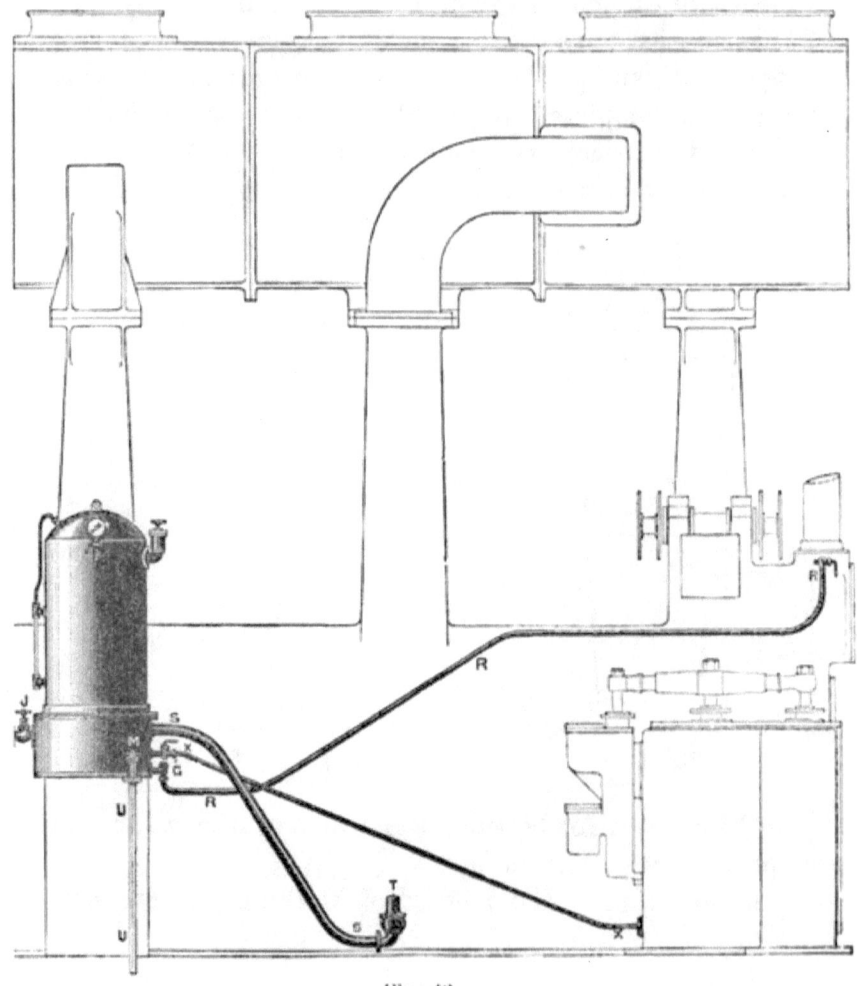

Fig. 43

the pipe S and reducing valve T. The necessary feed-water for the evaporator is taken from the circulating discharge by the pipe R, and enters the evaporator by the non-return valve G.

The steam condensed within the coils flows by the pipe X to the hot well. A pipe U conducts the brine from the brining cock M to the bilge. By placing the reducing valve T low on the condenser, as shown, the steam from the evaporator is allowed to mingle with the feed-water, and thus heat it to a slight extent. The amount of heat thus utilised is, however, necessarily small, as it is impossible to raise the temperature of the water above that corresponding to the vacuum in the condenser.

With this arrangement, as the water flows into the evaporator by its own weight, it is necessary to place the apparatus as low as possible, so that there may be a considerable head of water.

b. Evaporating into LP Casing and Condenser.

Fig. 44 shows an arrangement with connections to both LP valve casing and condenser. In this case it is necessary to fit a small pump V for the supply of water to the evaporator. This water is taken from the circulating discharge, as before, by the pipe R, and enters the evaporator by the pipe W and non-return valve G. A two-way cock Y is placed in the steam pipe S, and it is thus possible to discharge the steam from the evaporator into the LP valve casing, or into the condenser at will. A non-return valve Z should be placed on the LP valve casing.

In ordinary cases with the engine working, the steam

from the evaporator would be led into the LP valve casing; but when the evaporator is used in port the

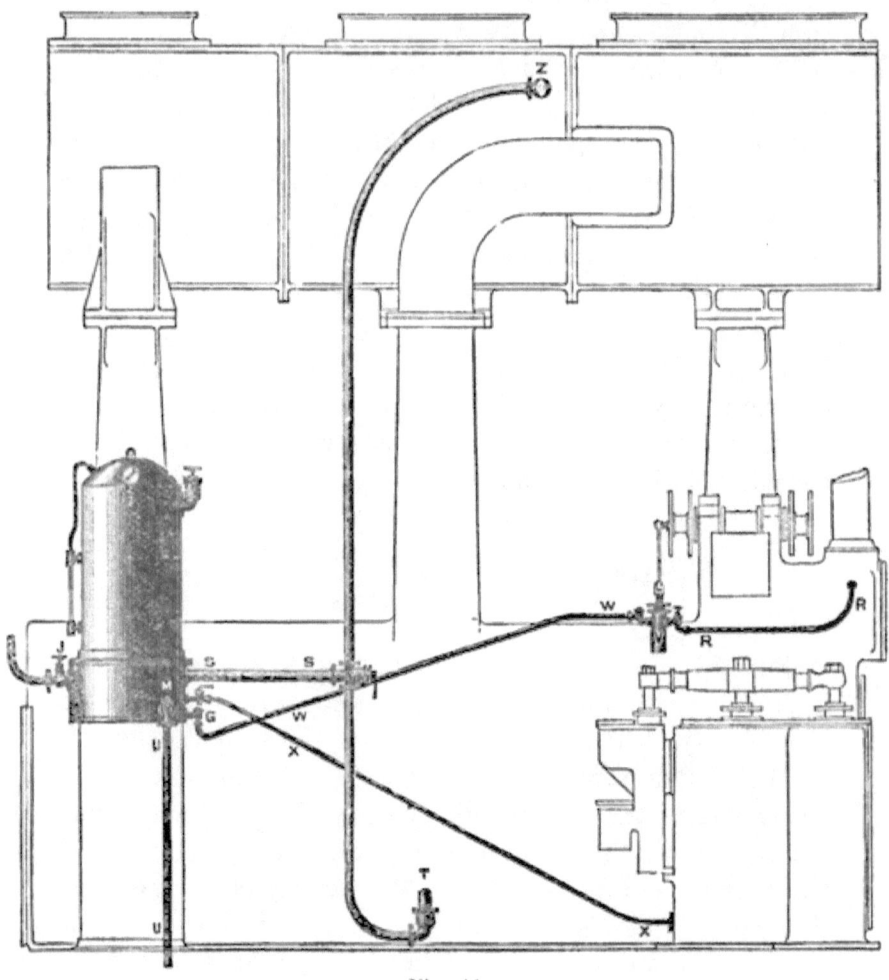

Fig. 44.

steam is discharged into the condenser, thus allowing any waste of water to be made up.

and Boilers and their Management. 105

c. Evaporating into Morison's Patent Feed-Heater.

Fig. 45 shows an arrangement with Morison's patent feed-heater and reducing valve.

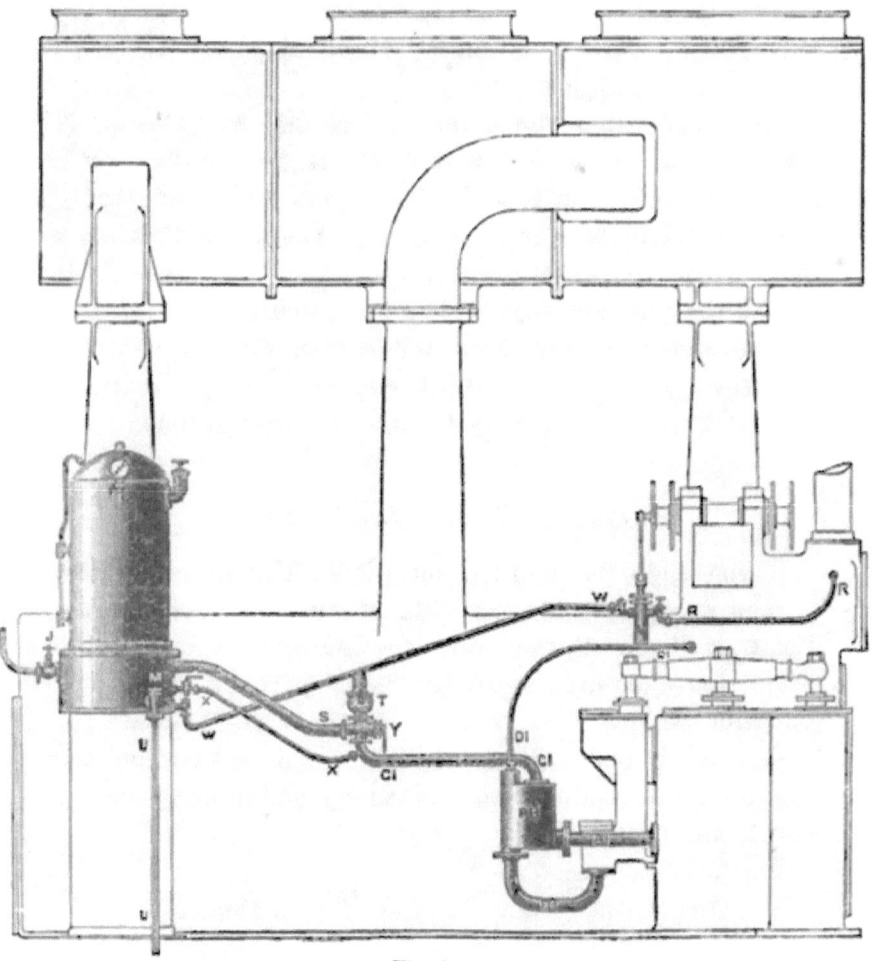

Fig. 45.

The feed-heater (which has been previously described)

is shown at FH. The water from the hot well enters by the pipe A1, and after being heated flows by the pipe B1 to the main feed-pump suction. A two-way cock Y is placed on the evaporator outlet pipe S, so that the steam from the evaporator may enter the condenser by the reducing valve T, or may be conducted by the pipe C1 to the feed-heater.

The drain from the coils may be led by the pipe X into the pipe C1 at any point beyond the two-way cock, as shown. D1 is a pipe for the escape of air and vapour from the feed-heater, and is led into the hot well at any point above the usual water level.

For the purpose of providing pure drinking water, part of the steam generated in the evaporator may be condensed in any fresh-water condenser, and, being absolutely free from grease, it is very suitable for this purpose.

BOARD OF TRADE REQUIREMENTS.

It may here be pointed out that when an evaporator is required for a passenger ship it must be made under Board of Trade Survey, and the material as well as the completed apparatus must be tested in the same way as a marine boiler.

Instead of the ordinary safety valve, shown on the engravings, a double lock-up safety valve with easing gear is provided.

The tests comprise :—

1. Testing the material at the steel makers'.
2. A hydraulic test at the manufacturers' works to twice the pressure in the main boilers.
3. A steam test on board ship.

In case of the steam test on board ship the Board of Trade requires the full pressure to be admitted to the coils (when connected to the boilers) and the evaporator safety valves to be tested for accumulation at 40 lbs. pressure.

Much trouble has been caused by evaporators being connected to reduced steam pipes, as the Board of Trade then insists upon the full boiler pressure being admitted to the coils during their test, and it is impossible to do this without interfering with the reducing valves.

As the Board of Trade does not recognise reducing valves however efficient, we strongly recommend that the evaporators be connected direct to the main boilers, or to some steam pipe in direct communication with the same.

LLOYDS' REQUIREMENTS.

Except when specially ordered the evaporators are made to pass Lloyds' requirements only, and are tested as follows:—

1. Coils and connections to twice the boiler pressure and to not less than 320 lbs. per square inch.
2. Dome and base to 80 lbs. per square inch.
3. Safety valve set to blow off at 10 lbs.

CHAPTER VI.

Combustion of Fuel—Economy of Cornish Boilers—How Combustion is Effected — Quantity of Air Required — Natural Draught — Forced Draught—Induced Draught—Different Methods of Effecting the Same—Their Respective Advantages and Disadvantages—Howden's System of Forced Draught—Results obtained from Actual Practice.

THE successive improvements in the marine steam engine during the last twenty or thirty years, whereby there has been effected so great a reduction in the ratio of fuel consumed to power obtained, have been almost exclusively confined to the motor, much to the neglect of the generator; the reduced consumption having been effected by a more judicious use of steam in the engine, and not by more economical evaporation in the boiler.

Putting aside the merely constructional improvements arising from greater experience, better material, and improved appliances, steam boilers for land and marine purposes are practically the same as they were at the time the average consumption of coal per horse-power developed in the engine was about three times that of the present day.

It may even be said that we have retrograded rather than advanced, because some forty years ago a higher evaporative economy was prevalent in Cornwall than is to be found there or anywhere else now. In that district, in 1846, an evaporation was recorded per lb. of coal of

12·89 lbs. of water from 212°. In the marine practice of the present day the average evaporation per lb. of coal is probably not more than ·6 of this Cornish economy, while in those high-speed steamships, which maintain an average of about 17 knots across the Atlantic, it is probable that barely half of this evaporative duty is reached.

This great falling off in evaporative economy in boilers of the present day must not, however, be ascribed to less knowledge but to the very different conditions and necessities that are now imposed by great industrial progress.

The conditions under which the high economy of the Cornish boiler was obtained are incompatible with marine requirements. They would also, as a rule, be impracticable or unprofitable if applied to land boilers, especially in populous districts.

Time and working space are every day growing more valuable. A maximum of power in a minimum of space is being more and more sought for. This demand is diametrically opposed to the essential conditions by which the high Cornish economy was obtained. The rate of combustion was from 3 to 4 lbs. of coal per hour per square foot of grate, accomplished by careful restriction of the air admission to the furnaces. The proportion of heating surface to water evaporated was from six to seven times that of the average in marine boilers. A comparatively low temperature in the high chimney gave an adequate draught for the slow combustion, and permitted the heating of the feed-water by the waste gases after leaving the boiler. Add to these precautions the most thorough protection of the whole boiler from loss of heat by radiation, and we have the conditions by which the high evaporative economy was effected in Cornwall forty years ago.

Though the necessities of our day render this cumbrous and slow system unsuitable for our steam supply, yet it must be acknowledged that the fact of the higher evaporative power now required from boilers being obtained only by a great sacrifice of evaporative economy is inconsistent with the general advance in engineering science, and contrasts most unfavourably with the marked economical improvements in the engine or motor.

This anomaly, however, is not incapable of being removed, and even a much higher evaporative power than now prevails may be had from boilers, along with a higher evaporative economy than is found in present practice; in short, an approach may be made to the unequalled economy of the Cornish slow combustion system with an evaporative power, it may be, ten times greater.

Experience has hitherto shown generally that with slow combustion judiciously carried out a high evaporative economy has been effected by natural draught, but as the rate of combustion has increased the evaporative economy has steadily fallen. Further, when combustion has been raised by forced blast to a rate much higher than is possible by natural draught, the evaporative economy has decreased in a still higher ratio.

It is important at this stage to elucidate the causes of this unsatisfactory result as far as possible before describing how the difficulty may be overcome.

In the case of a common fire-grate the fuel burns because the combustible matter it contains is brought into contact with that which will produce combustion, or, in other words, the carbon or hydrogen in the fuel is brought into contact with oxygen at a certain temperature, and to accomplish this either the gas or the fuel must be heated up to the temperature at which it will ignite with oxygen. Unless this is effected combustion, of course, will not take place; but when oxygen is brought into contact with carbon—heated as coal is in the fire—combustion will be continued until the whole of the fuel is consumed. It will be noticed that if coal is burned in an open grate the combustion is slow, and this is owing to what is called a bad draught. This cannot be due to an insufficiency of oxygen, because there is an unlimited supply of that gas available, and consequently we must seek for some other cause.

It would rather be due to too much oxygen, or, in ordinary language, too much air, in which case the cooling effect of the surrounding gas, oxygen and nitrogen, would prevent the temperature from rising high enough to admit of the rapid burning of the coal. If we take a fire in an ordinary open fire-grate, and close up the open part of it down to the level of the bottom of the bars, the fire will begin to burn up brightly, and the fuel will very soon become incandescent; in other words, its temperature will be raised very much indeed, the rate of combustion will be increased, accordingly there will be less visible smoke, and if used for raising steam such a fire would give a very much better supply than before. There has been no change in the draught due to the chimney, for it is the same chimney and there is the same area of fire-grate; but, in the present case, the supply of air has been limited, and the whole of it made to pass through the fire. As a natural consequence of decreasing the area through which the air can effect a passage its velocity will be increased, and the velocity with which the oxygen ordinarily meets the carbon particles will be so much increased that increased combustion and an intensified heat will be the result.

The ordinary fire-grate of a boiler is generally so arranged that nearly the whole of the air supplied to it has to pass under the grate-bars, no doubt a small portion is sometimes allowed to pass over the top of the fuel, but the greater portion has always to pass under the bars. Being applied to the under side of the fuel, and having to escape into the combustion chamber from the top side of it, the air is compelled to pass through the heated fuel, and in doing so it becomes thoroughly warmed, thereby bringing it into the condition most favourable for perfect combustion; and the small quantity admitted to the top of the fuel having likewise become similarly heated, a much better burning fire will be obtained than if it were open.

This is termed "natural draught," because there are no mechanical means adopted to produce it.

Combustion by natural draught in boilers at the rates now prevalent is necessarily more or less wasteful and imperfect, because the velocity given to the air supply, even with the sharpest draught, is insufficient to cause it to penetrate and mix with the fuel over the whole area of a furnace of ordinary dimensions, and further, because a very considerable portion of the heat of combustion is required to rarefy the air in the chimney to maintain the draught, this expenditure increasing with the rate of combustion.

Unless, therefore, a much greater supply of air is admitted to the furnace than is required theoretically for perfect combustion, it is found that, owing to an unequal distribution and imperfect mixture of the air with the fuel over the area of the furnace, part only of the oxygen enters into union with the combustible gases and carbon of the fuel, the remainder passing to the chimney unassimilated. The unavoidable consequence is that a considerable portion of the carbon leaves the boiler unconsumed as carbonic oxide. Visible smoke is also formed in large quantities, and though this objectionable product may contain but a small portion of carbon, it is always an indication of imperfect combustion and of the waste of more important constituents of the fuel.

On the other hand, if an increased admission of air is made to the furnace, in order to secure the more complete combustion of the fuel, then the excess of air—which must always be largely beyond the quantity required for perfect combustion when properly carried out—reduces the temperature of the furnace, and consequently the ratio of evaporative power, and, what is still more wasteful, robs and carries off to the chimney a large percentage of the total heat of the fuel.

The waste arising from imperfect combustion caused by restricting the air supply, as described, may be estimated from the following data. When perfect combustion is effected by the chemical union of two atoms of oxygen with one of carbon forming carbonic acid, 14,500 units of heat are evolved from 1 lb. of carbon; but when from imperfect combustion only one atom of oxygen unites with one atom of carbon and carbonic oxide is formed, only 4400 units of available heat are evolved from 1 lb. of carbon. The loss arising from this imperfect combustion can, therefore, be calculated by ascertaining the proportion of carbonic acid to carbonic oxide in the waste gases, noting also the quantity of air passing away with its oxygen unassimilated.

The waste arising from admitting an excess of air may be estimated by ascertaining the quantity of air admitted in excess. About 11 tons of air are required for the perfect combustion of 1 ton of ordinary coal, if no portion of the air is wasted. Double this quantity or even a larger proportion is frequently supplied. If double the quantity be assumed, then 11 tons of superfluous air must be heated from the combustion of the 1 ton of coal, which will not only greatly reduce the temperature of the furnace and its evaporative power, but in leaving the boiler at a temperature, it may be $600°$ or more, higher than it entered the furnace, will also carry off a large percentage of the heat of the fuel. The loss caused by 11 tons of superfluous air leaving the boiler at $600°$ above the entering temperature stated in figures is $11 \times 2240 \times \cdot 246$ (the specific heat of the escaping gases) $= 6061 \cdot 4$, which, $\times 600 = 4,638,840$, the loss in units of heat, being fully $12\frac{1}{2}$ per cent. of the total heat of combustion of the ton of coal, taking the efficiency of the coal at 12,600 units per lb.

The dilemma, therefore, in which every furnace operator finds himself when working with natural draught is the choice of either admitting a moderate amount of air to the furnace, and having in consequence an imperfect combustion, with much smoke and considerable waste of fuel, or of admitting air greatly in excess of what is required for perfect combustion properly effected, and thereby obtaining a more complete combustion, but at the cost of a large waste of heat carried off to the chimney by the superfluous air, besides the reduction of the evaporative power of the boiler by the dilution of the furnace temperature.

Of these two sources of loss, unfortunately, the greater is generally that arising from the too liberal supply of air, consequently the hotter furnace, with increased steam and much smoke, is preferred by all stokers to the cooler furnace with diminished steam and little smoke.

The well-known disadvantages, such as enumerated, attending the combustion of coal by natural draught, with the limitation of the supply of steam, often further affected by the state of the draught and quality of coal, have led to frequent attempts to effect combustion in boiler furnaces by air supplied by mechanical means above atmospheric pressure, that is, by means of forced draught.

To make the meaning of this phrase more intelligible, it may be defined as follows:—When any mechanical appliances are fitted to the front of the fire for the purpose of forcing air into it at a greater velocity than that it would acquire from the chimney alone, then such an arrangement is termed "forced draught."

In any case the object of such an appliance is to increase the supply of air to the furnaces, and to do it in such a way that it is applied wholly to the combustion of the coal, and not to the dilution of the products of combustion, which would lower the temperature both above the fire and in the combustion chamber and tubes. If any appliance used for this purpose produced such a result it would be very bad indeed, for most of the smoke-consuming apparatus are dangerous on that account.

As already stated, the main object of all such appliances is to increase combustion, and that increased combustion is, of course, effected upon the same area of fire-grate. For instance, take the case of a boiler with 30 square feet of fire-grate, it burns 14 lbs. of coal per square foot per hour, and a certain quantity of steam is produced,—now, if the combustion can be so increased as to burn 21 lbs. of coal per square foot, 50 per cent. more steam would be expected.

Some sixty years ago, when marine engineering was in its infancy, the funnels of steamships were of small size compared with those now in use; it is not to be wondered at, therefore, if the consumption was not very great in the furnaces. They burnt then about $8\frac{1}{2}$ lbs. of coal per indicated horse-power per hour, which meant, of course, a very large consumption, compared with what is the case now, so that a comparatively small steamer would burn a large quantity of fuel. Considering that ships then had comparatively little depth of hold, and were burning a large quantity of fuel per indicated horse-power, it may be imagined that attention would sooner or later be drawn to the necessity for something more than a funnel.

Coming to more modern times, or about twenty years ago, from 14 to 20 lbs. of coal per square foot of fire-grate was the usual consumption; in fact, 14 lbs. was considered a very fair average rate for ordinary sea-going ships with average stokers. On trial trips, with good fires, good coals, and good stokers, as much as 20 lbs. were sometimes burnt. This increase in consumption of coal per square foot of fire-grate was due to another cause operating, and that cause showed its effect pretty plainly later on when steel was introduced, and furnaces were made very much larger in diameter than formerly. When this had been done, grates were consequently decreased in length in proportion to their increased breadth. This change came about slowly without any one noticing the effect, and the only indication of it was when it became known that some engineer, by shortening his furnace bars, had obtained better results than formerly. It seemed very much like a paradox, but there appears very good reason for it on mature reflection, for it will be readily seen that any increase of grate surface obtained by an increase in diameter of furnace must be small in comparison with that effected in the area for air supply, because, while the grate area would increase directly with the diameter of the furnace, the area for air supply would increase in the ratio of the square of the diameter.

When from special circumstances, therefore, larger furnaces could be put in, better results were, of course, obtained. One of the causes which first operated against this improvement was the comparatively high price of Low Moor iron which, in the case of large plates, amounted to as much as £40 per ton. This was a great consideration, but with the advent of steel there became practically no limit to the size of the plate. After this came the Bowling hoop, followed by Fox's corrugated and other patent furnaces, so that it may be said there was then practically no limit to the diameter of furnaces.

It will be seen that at the outset improved combustion was obtained from increase of air supply with decrease in length of bar, for the fire-bars were actually made shorter. In actual work, with long grates, a considerable portion is ineffective, for a fireman cannot work a grate very efficiently if it is much over 5 feet 6 inches long, while, if he has a grate 4 feet 6 inches long, the back end is kept in as good order as the front and as well supplied with coal.

Another means which tended to the same end was an improved form of fire-bar. There are several patterns of these, such as Martin's, Henderson's, and others, by the use of which the fires are always kept clean, and the efficiency of the whole of the air spaces between the bars is maintained with, of course, a very much better result. Probably with Henderson's bars, used at a fairly good speed, nearly as good results might be got as with moderate forced draught; these bars may themselves be considered a mechanical means of improving combustion, but they do not come up to the strict definition of the term, inasmuch as they do not produce of themselves any larger inflow of air to the furnaces.

The lengthening of the funnel is, of course, another method whereby an improved draught may be produced, and every one who has to do with boilers is supposed to be familiar with the fact that an addition of a few feet to a funnel is relied upon as one of the readiest means for improving the efficiency of a boiler.

The repeated attempts to utilise forced draught, however, have not hitherto been successful, and therefore, except in locomotives and torpedo boats, natural draught is still in general use for steam boilers.

The causes of failure though various, and often not at

first sight apparent, become, nevertheless, quite obvious when sufficiently investigated, and a short description of the most prominent of these attempts to improve combustion by mechanical means may be of some interest.

The plan which has probably been most frequently tried is that of forcing air by a fan or other blower into a closed ash-pit, but this mode of increasing the power of a furnace, though apparently simple, must always be disappointing, for the following reasons:—

If the coal on the fire-grate be kept sufficiently thin to permit the oxygen of the air supply to penetrate to the upper layers of the fuel, to effect combustion of the inflammable gases, the quantity of air passing through will be too great, and the temperature of the furnace will be uunecessarily reduced, and much loss of fuel occur.

If, on the other hand, the coal be laid sufficiently thick to prevent an undue quantity of air from passing through, there will be a lack of air to consume the combustible gases generated by the great heat from the combustion of the lower layers of fuel. Much visible smoke will in consequence be produced, and also carbonic oxide, so that in this case also less economy will follow than may be realised by natural draught. This system of supplying air to a furnace is likewise very severe on the fire-bars, and has other disadvantages, such as requiring to shut off the air when the furnace door had to be opened. If this system is worked so that the air is prevented from passing through the fire-bars, and fuel sufficiently to produce any pressure in the furnace by reducing the air spaces between the bars, and the chimney draught therefore capable of drawing air by the furnace door, some of the difficulties mentioned will disappear; but in such a case, though the air may be supplied under the bars by a fan with pressure, it cannot be termed an example of "forced draught," as it does not comply with the terms of our definition.

In 1862 a boiler was constructed by Mr. Howden for the purpose of testing the forced draught method of effecting combustion by air under pressure. The boiler had a short chimney, and the air was supplied to the closed ash-pit by an engine and fan. The results of the trials were disappointing, and exactly of the character already described; but they were not altogether valueless, for they clearly showed that forced combustion could not satisfactorily be carried out under the conditions existing at these trials.

A second method resorted to for increasing combustion in boiler furnaces is that of exhausting the air in the chimney or uptake by a fan, thus reducing the pressure in the flues or tubes and furnaces, and thereby inducing a more rapid current of air through the furnace, both below and above the grate-bars. This plan, which is known as "induced draught," so far as the supply of air to the furnace is concerned, is in practice more workable than the plan previously described, as the air enters the furnace not only at a greater velocity than is attainable by natural draught, and is more thoroughly intermingled with the fuel all over the furnace, but, the pressures above and below the grate-bars being equal, the operations of the furnace become much more easily managed.

This mode of creating these advantageous conditions in the furnace is, however, so far objectionable. The passing of the hot gases of combustion through a working fan is of itself a mistake, practically and theoretically. Even if it were possible for the machine to continue in working order under this ordeal for any length of time, the hot gases, if leaving the boiler at no more than 461° above the entering temperature, would be twice the volume of the air which entered the furnace from the stokehold. A fan to exhaust the hot air would therefore require to be at least double the capacity of one which would be able to supply the same quantity direct from the stokehold to the furnace. This plan, which was one of the earliest tried, is therefore considered impracticable for large boilers, but has been occasionally used in boilers of a limited size.

A third method, producing the same effect on the furnace as the last, is that of inducing a draught by means of a steam jet in the funnel. This plan is simple, requires no working machinery, and is within certain limits effective. It illustrates the advantage of mixing the air more thoroughly with the fuel than is possible by natural draught, as by its use a greater evaporative power is obtained, while the consumption of fuel per horse-power of the engine is not more than the rate per horse-power for natural draught, notwithstanding the loss of steam by the jet.

The waste of water by this system unfits it for marine boilers working at high pressure, and it is not required in land boilers with high chimneys. It is, however, seen on a grand scale and in its most legitimate use in the locomotive boiler, where, instead of merely a small percentage, the whole of the steam from the boiler is used for the purpose of inducing a current through the furnace, after it has done its work in the cylinders.

The remarkable effect produced by the powerful action of the blast-pipe is well known. It has increased the power of the boiler to at least six times the efficiency obtainable by natural draught; and what we owe to this power of increased combustion is simply incalculable. It is not too much to say that the amount of travelling and carriage of goods from end to end of the kingdom could not have reached a tithe of their present enormous extent but for this powerful artificial supply of air to the furnace of the locomotive boiler. An average consumption of 100 lbs. of coal per square foot per hour is burnt on the fire-grate of an express locomotive, and, what should not be overlooked in the consideration of this subject of combustion under air-pressure, this astonishing rate of combustion is accomplished with comparative ease and remarkable economy.

A further method of improving combustion, and more especially of preventing smoke, which has been frequently tried, is that of increasing the supply of air to the fuel above the grate by currents induced by steam jets. The steam jets penetrate the fuel, carrying with them the air to supply the necessary oxygen to those parts of the furnace which would not be sufficiently reached by natural draught, and, by this means, combustion is improved, and smoke in large measure prevented.

This is not in itself an economical plan; it has several objectionable features; and for marine boilers it is, for obvious reasons, quite unsuitable. It gives another proof, however, of the advantage, within certain limits, of an artificial supply of air above atmospheric pressure, as the improved combustion effected compensates to a considerable extent for its inherent wastefulness.

The most successful plan, in some respects, that has hitherto been applied to marine boilers, is that of the air-pressure stokehold or boiler-room, as used in torpedo boats, in which the air enters the furnace above and below the fire-bars in the same manner as by natural draught, but at the increased velocity due to the difference of air-pressure in the boiler-room over that of the atmosphere outside.

The results obtained in evaporative power approach those of the locomotive boiler, and the comparatively small boilers used in these boats, generally of the locomotive type, under the powerful action of the great supply of air to the furnace under pressure, are able to generate steam sufficient to impart a speed of 25 or 26 miles an hour to the vessel.

The success of this system in torpedo boats led our own and other governments to apply it to large steamers having a number of boilers, with several furnaces in each; but, it being impracticable to make air-tight a large boiler compartment, and the coal bunkers which must necessarily be in open communication therewith, the plan was adopted of forming air-tight chambers, or rooms, on the furnace ends of each boiler, or pair of boilers, into which the air can be forced by fans working on the deck above. These rooms contain the coal being used by the stokers on duty, and access to the chambers is obtained by doors specially designed to prevent egress of air.

The results of a number of trials of this arrangement showed that much larger supplies of steam could be obtained by it, and with much less discomfort to the stokers than would have been possible from natural draught.

This, however, is about all that can be said in favour of this system. Its practical disadvantages are very considerable, even in war vessels, and are more than sufficient to prevent its application to mercantile steamers. The chief defects of the system, and they are serious, are its wastefulness and its tendency to injure the boilers. It is a system which may be characterised as one for obtaining an increase of power regardless of expense. Judging from the experiments already described, the impossibility of working a furnace with air under pressure without waste of fuel, unless most carefully studied precautions are adopted, it may readily be seen that the indiscriminate admission of air in this system must be wasteful of fuel in a very high degree. The effect on the boilers themselves, whenever a furnace door is opened, of the rush at full pressure of a column of cold air, equal to the area of the furnace door, at a velocity from 80 feet to 100 feet per second, must be most injurious. The furnace plates, as well as the tube and other plates of the combustion chamber, cannot but suffer severely, while the great volume of cold air lowers the temperature of the whole boiler, and greatly neutralises the benefit of the rapid combustion of fuel; and when scale is formed on the plates to any extent, the injury from this mode of working becomes very marked.

In treating of this subject, we cannot do better than begin by describing an arrangement devised by Mr. Howden during the year 1880, to which he was led by the remembrance of the causes of failure to obtain satisfactory results in 1862, as already narrated, and by observing the defects of the other methods previously referred to.

In the summer of 1882 he had an opportunity of applying it to a small marine boiler, 7 feet diameter by 7 feet 9 inches in length, with two furnaces, each of 2 feet diameter, and the results being satisfactory he thereafter constructed a special boiler for the purpose of testing, on a sufficiently large scale, the material advantages anticipated from the adoption of this mode of working. This boiler was 10 feet in diameter by 9 feet in length, with two furnaces each 3 feet inside diameter; the furnace tubes extended the whole length of the boiler, the fire gases returning to the front from a dry chamber at the back end lined with fire-brick and otherwise protected from radiation of heat, through two separate stacks of tubes, each stack having 45 tubes $3\frac{1}{4}$ inches external diameter.

This boiler was erected in a yard by itself, with evaporating tank, engine and fan, and other appliances suitable for testing results, and has been used more or less frequently since then in making trials under various conditions.

The principal objects which it was endeavoured to accomplish satisfactorily were :—

1. To effect with ease and certainty complete combustion of fuel in the furnaces of ordinary steam boilers, at all rates from zero up to that of a locomotive boiler if required.

2. To effect this combustion economically as regards fuel, at all rates, whether high or low.

3. To effect this economical combustion rapidly or slowly in boilers of ordinary character, either singly or in numbers, without any change in the usual conditions of stokeholds when boilers are worked by natural draught.

4. To effect the foregoing purposes with less wear and tear of the boilers and furnace fittings, and with less discomfort to the stokers and attendants than is experienced when working in the ordinary way with natural draught.

The problem set forth under the first head made it necessary to use air under a pressure above the atmosphere, such as is conveniently got from a fan, with simple means of using more or less of this air as required.

To attain the second point, that of economical combustion at all rates, it was imperative that the supply of air should not only be under perfect control, but that the necessarily varied quantities admitted per unit of time or of coals consumed should be exactly ascertained; and, further, its distribution had to be as nearly as possible equal over the furnace, and limited as nearly as possible to the theoretical quantity for the weight of fuel being consumed; and, finally, that the waste of the heat of combustion should be reduced to a minimum.

To accomplish the third, it was necessary that the supply of air to any single furnace in a boiler, or to any boiler in a series, should be quite independent of the supply to the other furnaces or boilers.

The advantages under the fourth head follow from the more economical combustion, requiring less supply of fuel than would be necessary with natural draught, by the radiation of heat into the stokehold from the furnaces being prevented, and by the supply of air to the fan being made to circulate through the stokehold. The saving of tear and wear is due to the air being introduced to the furnace so as to preserve the furnace fronts and interior fittings, and also to the prevention of the usual injurious rush of cold air into the furnace when stoking, by suspending the air admission when a furnace door is opened.

The means by which it was proposed to carry out these several purposes will be understood more clearly by the following description and reference to the drawing.

On the front end of the boiler an air-tight chamber or reservoir (Fig. 46) is formed of light plate iron, to receive the air for combustion under pressure from a fan. This air chamber extends over the whole front of the boiler, from some distance above the upper row of tubes downwards, so as to enclose the furnaces and ash-pits. The smoke-boxes of each stack of tubes are completely separated from the air chamber by suitable castings, as are also the furnaces and ash-pits, so that no air can enter the furnaces or ash-pits except through passages regulated from the outside by valves.

The upper portion of the reservoir extends outwards from the boiler not more than the ordinary smoke-boxes and uptake; below, at the furnaces, the reservoir in this case extends $7\frac{1}{2}$ inches from the boiler, but this may be reduced.

On the front of the castings which separate the furnaces from the air chamber, the outer furnace doors and ash-pit doors are hinged air-tight. Attached to the outer furnace doors by studs, and moving on the same hinges, are the inner and proper doors of the furnaces, which shut on the inner front plate in the usual manner.

A dead plate between the outer and inner front plates separates the ash-pits from the furnaces above the fire-bars. Into these spaces above the dead plate and be-

128 *Triple and Quadruple Expansion Engines*

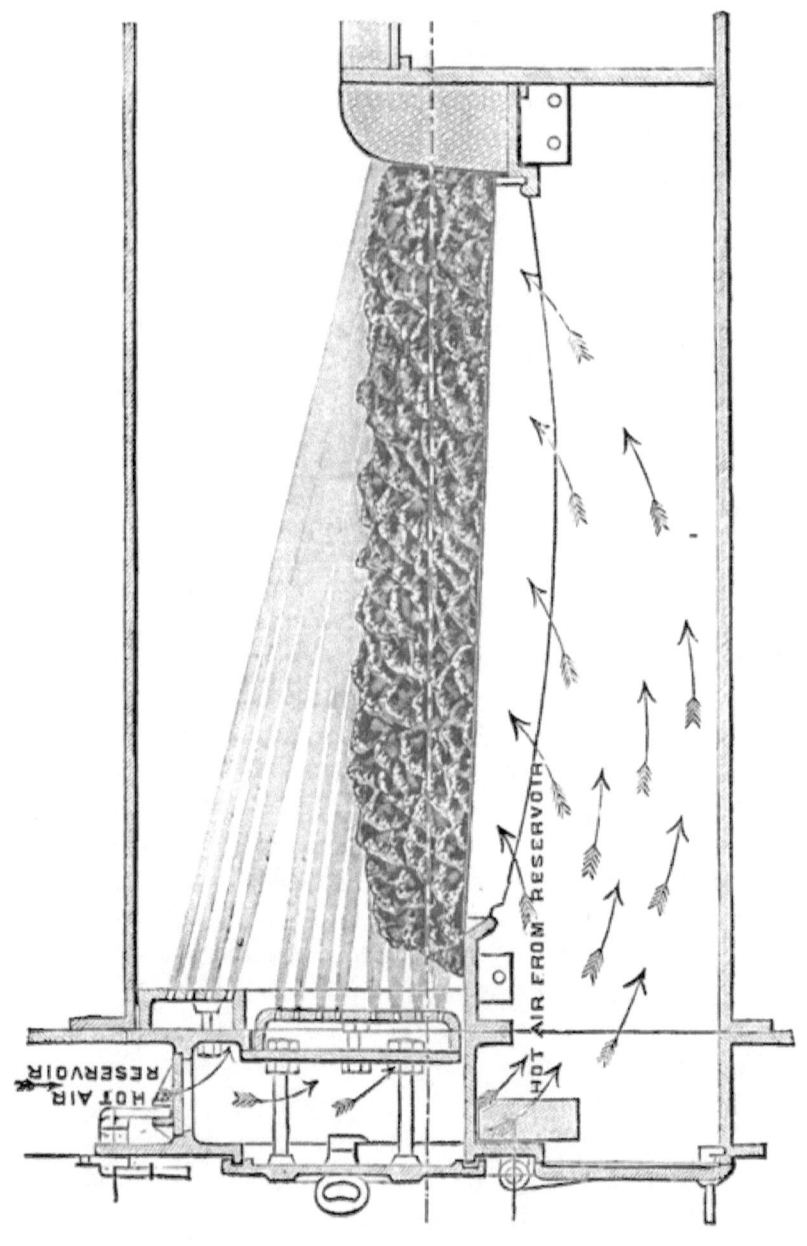

tween the outer and inner furnace doors the air is admitted in the desired quantity from the common chamber by simple plate valves, and after passing through perforations in the doors, at the sides, and above the doors, is received into cast-iron boxes, which serve the double purpose of protecting the front plate and furnace door from the great heat, and distributing the air effectively among the fuel. The air is admitted as required to the ash-pits by separate valves.

By these arrangements the admission of the air to different parts of the furnace is kept under perfect control, the quantities admitted being ascertainable, and all radiation of heat from the furnaces or ash-pits prevented.

The heat of the fire gases after they leave the boiler is largely utilised by causing them to pass up through the tubes in the air reservoir immediately above the smoke-boxes. While the hot gases pass through the inside of these tubes, the cold air from the fan entering by the pipe is made to sweep right and left among the tubes, carrying a considerable portion of the escaping heat downwards by the sides of the smoke-boxes to the furnaces.

This heating of the air for combustion by the waste gases is an element of the first importance in effecting a high economy of fuel in forced combustion, and the one which makes possible or increases the beneficial effects of other combining elements. The various beneficial effects of the heated air are not to be measured only by the direct recovery of so much heat that otherwise would be lost, but also by their effect in increasing the average temperature of the furnace, which again not only adds to the evaporative efficiency of the boiler, but also allows of a more facile and rapid union of the oxygen of the air with the gaseous products and the carbon of the fuel, so that actually *less weight* of hot air than of cold is required to effect the combustion of an equal weight of fuel in a given time. This further raises the degree of evaporative economy and power of the furnace, making an economy and rate of combustion possible which otherwise would be unattainable.

When a furnace door is opened for stoking purposes, the air admission valves for the ash-pit are closed, so that the furnace may be operated on in a quiescent state. No rush of cold air can thus pass into the furnace, as in the air-pressure stokehold system, or even as in boilers worked by natural draught. The saving of the boiler from injury, and the prevention of waste of fuel, by this means are important features.

When it is desired to reduce or increase the rate of combustion, the admission valves are opened or shut as required, and with any given pressure of air in the chamber it is not difficult, after a few trials, to establish the required openings for different rates of combustion, from zero to the highest rate possible by the given air-pressure.

When the admission valves are entirely shut the combustion may be practically suspended for hours, the fires continuing more or less incandescent.

By this means a steamer may with ease be kept lying under any desired pressure of steam for any length of time, and without tendency to blow off.

In erecting this boiler for trial, Mr. Howden purposely made the conditions under which it was to work as difficult in regard to combustion as would ever be likely to occur under the most unusual circumstances in actual practice. The funnel was made only 15 inches in diameter, and 21 feet high from the fire-grate. A still greater restriction was the limiting of the air-heating tubes in the reservoir (through which the whole of the fire gases issuing from each furnace have to pass to the common uptake) to 14 in number, of $3\frac{1}{2}$ inches external diameter. The effect of these restrictions and contractions is to reduce largely the rate of combustion with a given pressure of air.

With natural draught this same boiler generates steam very slowly; but with 2 inches to $2\frac{1}{4}$ inches of air-pressure a combustion of about 30 lbs. per hour per square foot of fire-grate can be reached, and this rate could evidently be doubled with a corresponding increase in outlet area, or by increasing the air-pressure, but the latter would be undesirable where the former can be accomplished.

It may be added that, under the restricted conditions described, the boiler was worked with the utmost ease at any rate of combustion up to the limit mentioned, and with an absolute freedom from visible smoke.

In ascertaining the effect of the admission of air above, as well as under the fire-grate, trial was made with different areas of apertures, and of different pressures of air on equal areas. For example, the trials were begun with a limited aggregate area in the holes of the air-boxes inside the furnace, this area being gradually increased until it was found that the increased area added little to the rate of combustion, while the evaporative economy became sensibly reduced.

The admission below the grate was also tried with various areas of air inlet apertures and pressures, and their effect noted on the combustion and evaporation.

In the course of these trials some interesting results were noticed. For example, when the air admissions above and below the fire-bars were regulated in certain proportions to each other, if the admission openings above were then decreased and those below increased to about the same amount, a much higher rate of combustion took place than when the process was reversed by increased admission above and decreased below; this difference arising from the changes produced in the balance of pressure in the furnace.

When the pressure was relieved somewhat above and increased below, the air passed much more rapidly through the fuel in its passage to the chimney, with a correspondingly more rapid combustion, than it did when, with the same aggregate admission, the pressure was increased above and decreased below; and in this case the current upwards through the fuel was greatly checked, and matters could be so regulated that the current through the fuel could be nearly prevented by the pressure above.

At the same time the admission of air above the fuel in proper proportions serves most important purposes, and contributes towards realising the highest economy. It has also an important effect in preserving the furnace-fittings and fire-bars. The great destruction of fire-bars caused by working with the closed ash-pit system has been already referred to; and the cause of this destruction became evident after several trials with this same boiler, as in it the admission above the bars tempered the velocity of the air through the fuel from below, while the fuel was kept in perfect combustion above. When the pressure of air is entirely from below, as in the closed ash-pit system, with a sufficient depth of fuel on the grate, and a given air-pressure, the bars can be melted by the intense heat generated in the lower layers of the fuel; but in the experimental boiler the fire-bars and air-boxes in the boiler were found to be, on examination, still sharp on the corners, and as good as when put in, after making all the trials mentioned, and consuming in some cases about 30 lbs. of coal per square foot per hour. This of itself is an important practical result in the working of furnaces with air under pressure.

The chief point, economically, towards which Mr. Howden had been working in these trials, was the attainment of a high rate of combustion with a minimum admission of air, that is, an admission as near to the theoretical limit as can be practically reached. This is no easy matter, and requires to be approached step by step. The trials would have been in great measure a groping in the dark, even with the appliances described, but for the invention of an apparatus for ascertaining by measure the quantity of air entering at different parts of the furnace both above and below the fire-bars at the various pressures used. This device is the application of an anemometer, with an apparatus in which the same conditions as to entering and back pressures are created as those which exist in the part of the furnace into which the air is being admitted. The velocity is read off without difficulty, and the dimensions of apertures being known, and temperature ascertained, the volume of air is exactly calculable.

This system of air measurement, combined with chemical analysis of the products of combustion, enabled the results to be ascertained with intelligence and certainty. These trials have unquestionably shown the great difficulty of obtaining a high rate of combustion with forced draught, without admitting a most wasteful excess of air; and that, although it is comparatively easy to obtain a high rate of combustion, it is very difficult to combine it with economy, unless suitable adaptations are used for that purpose.

In the small boiler, which was first tried with a very complete combustion, the evaporation per lb. of coal was very low, entirely owing to the excess of air. In the special boiler the evaporation has already reached a very fair economy, though not within a considerable distance of what is yet almost certain to be attained.

What has already been accomplished shows that from $9\frac{1}{2}$ to 10 lbs. of water, at 212°, could be evaporated in boilers at sea, from 1 lb. of Scotch coal, with a rate of combustion of 30 lbs. per square foot of grate per hour; but there are good grounds for expecting that an evaporation of even 12 lbs. may yet be reached with a rate of combustion from 40 to 50 lbs. per hour per square foot of grate.

It should be explained that the boiler used for these trials, with the furnace arrangements as shown, was not intended to represent the best mode of carrying out this system of combustion, it was merely a boiler suitable for ascertaining results under varying conditions. There were several changes afterwards made in its fittings in the light of the results already obtained, which will, without doubt, increase the evaporative economy without reducing the rate of combustion.

Compared with the working of boilers by natural draught, the advantages which this system of combustion by air under pressure gives may be summed up as follows :—

1. Complete combustion of fuel of all qualities, under conditions in which combustion could not efficiently be obtained by natural draught.

2. The power of regulating with ease the amount of combustion desired from zero to many times that possible by natural draught; also the capability of maintaining the fuel in the furnace incandescent for a considerable time without appreciable consumption.

3. A great reduction in the size or number of boilers required to produce a given power, and the capability of increasing the power in steamships far beyond that now attainable with boilers worked by natural draught.

4. Greater economy in producing steam from the following causes :—

a. From more complete combustion of fuel than is attainable by natural draught with a reduced admission of air.

b. From the higher temperature of the furnace arising from the more perfect and higher rate of combustion, and from the air supply being partially heated before entering the furnace.

c. From the utilisation of the waste heat of the escaping gases.

d. From the prevention of heat from the furnaces and ash-pits being radiated into the stokehold.

e. From the much less expenditure required to supply

the air of combustion from a fan than is required to heat a column of air in a chimney to obtain supply by natural draught.

f. From preventing the cooling down of the boiler by a rush of cold air to the furnace when a furnace door is opened.

5. Less discomfort in stoking, the stokehold being kept fresh and cool by the radiation of heat from the furnaces being prevented, and the fan drawing fresh air into it continuously, independently of ventilators.

6. The complete absence of the great nuisance of smoke in the use of steam-power.

7. The abolition of all unsightly chimneys in town and country now necessary for combustion by natural draught.

Some engineers have expressed an opinion adverse to the practice of heating the air of combustion by the waste gases instead of the feed-water; estimating the abstractive powers of water and air to be as the products of their specific gravities and specific heats, thus making water 3500 times more efficient than air; but this is altogether an erroneous conclusion, as the success of Howden's system is largely due to the fact that air is a greatly better abstractor of heat than water passing through a feed-heater; and the relative values in this respect may be obtained from actual trial sufficiently correct for all practical purposes.

In the *New York City*, for instance, the heat abstracted by the air for combustion from the escaping gases in the air-heater, on ordinary sea working, averaged $190°$. Taking the air used for combustion at $18\frac{1}{2}$ lbs. per lb. of coal, we have $190° \times 18·5 \times ·238 = 836·57$ units of heat directly recovered by this air from the waste gases per lb. of coal consumed.

In a boiler lately tried, where the whole of the escaping gases passed through a horizontal multitubular feed-heater, and where the water, evaporated from 42°, was 10 lbs. per lb. of coal consumed, the temperature of the feed-water, leaving the injector at 114°, was raised, on an average, only 25° in passing through the feed-heater, or $10 \times 25° = 250$ units of heat recovered by the feed-water per lb. of coal consumed.

This does not, however, exhaust the case in favour of air as a better heat abstractor, for the feed-heater referred to had a tube surface of 304 square feet, with only 75 lbs. of water passing through it per minute at a velocity of ·025 feet per second, while the *New York City's* air-heater has only 230 square feet of surface, with an average of 275 lbs. of air passing through it per minute at the velocity of 25 feet per second.

Before adverting to the important elements of proportioning the quantities of air for combustion and the proper manner of bringing the air in contact with the fuel, we may refer to some vague ideas and curious misapprehensions which appear common in regard to such points.

In some recent papers bearing on the subject, it appears to be taken for granted that all that is required to secure certain temperatures in, and results from, a furnace, is the supply of a given weight of air for a given weight of combustible, the resulting temperatures being in an inverse ratio to the weight of air supplied. In one of these its author, founding his argument on a table in Dr. Rankine's *Manual of the Steam Engine*, goes on to say—"If the temperature of the atmosphere is 60°, and the fuel is burned by just so much air as contains the necessary amount of oxygen for combustion, viz. 12 lbs. of air per lb. of fuel, the resulting temperature of the products of combustion will be 14°·640. If 18 lbs. of air are used the temperature will be 13°·275, and if 24 lbs. of air, the temperature will be 12°·500." In a similar paper, minute calculations were worked out from similar hypothetical and theoretical bases, of results expected from boilers with given proportions of grate and heating surface; but calculations of furnace temperatures and effects on such bases are misleading and delusive.

The temperatures taken from Dr. Rankine's *Manual* are merely the theoretical values of the combustion of pure carbon at the instant when combustion is complete, on the hypothesis that the proportions of air stated could be supplied without any modifying elements intervening, —conditions impossible in furnaces,—and were not intended by Dr. Rankine to represent actual temperatures in furnaces. It is quite gratuitous advice for any one to say use forced draught with certain proportions of air to fuel; or use in ordinary combustion certain proportions of air to fuel, and certain results will follow, without such an one showing how the things recommended can be done in practice.

In the furnace of a steam boiler, worked in the ordinary way with the respective air admissions mentioned in the paper, the relative resulting temperatures would actually be reversed. If 12 lbs. of air per lb. of fuel were introduced into a boiler furnace in operation in the ordinary manner, a much lower temperature than is usual in boiler furnaces, and a more wasteful combustion would, for obvious reasons, result.

If 18 lbs. of air per lb. of fuel were admitted in an equal time to the same furnace, a considerably higher temperature and less wasteful combustion would follow, and with 24 lbs. of air both temperature and economy would be higher still.

This would lead to the conclusion that 24 lbs. of air admission would give the best result, from the fact that 24 lbs. of air admission appears to be about the average in well-managed furnaces worked in the ordinary manner.

Now, the reason of this is not because such a large proportion of air is desirable, but because it is necessary in furnaces worked in the ordinary manner, for if the same active combustion could be effected in the same time with less air, a proportional increase of temperature and economy in fuel would inevitably follow. Universal experience, however, shows that to obtain a certain evaporative power from a boiler worked in the ordinary way, about 24 lbs. of air must be used. Had it been possible, by merely shutting off a certain quantity of air from a furnace, to have obtained a higher temperature and better results, it would have been discovered long ago. The apparent anomaly—of the worst practice, theoretically, giving the better result—arises from practical conditions due to the manner in which the air is admitted to the furnaces, and the behaviour of the gases in combustion.

Temperature in furnaces does not depend primarily on the weight of air used per lb. of fuel consumed, even when that combustion is judiciously effected, but on the quantity of fuel brought under combustion in a given time and space, the greater the quantity consumed the higher the temperature.

Returning to the consideration of the effect of the manner in which fuel is burned, it may be pointed out that though it is impossible in the large scale of a boiler furnace to reduce the proportion of air very closely to the theoretical quantity sufficient for the complete combustion of the fuel, yet a very large reduction can be made on the proportion now used by a different mode of making admission. Except for the hydro-carbon gases, which apparently cannot be consumed without a considerable excess of oxygen, very little excess would be required for the complete combustion of the solid carbon by the adoption of effective means for combination. A large dilution of the carbonic acid in the furnace by admission of excess of air is only necessary when furnaces are worked in the ordinary manner. What is wanted is to bring the air for combustion *simultaneously* in contact with the gaseous and solid fuel *over the whole surface of the furnace* at a velocity that will insure its intimate mixture directly with the fuel. In such circumstances a greatly reduced proportion of air will suffice for combustion, more especially if the mode of effecting combustion is at the same time so arranged as to gasify the carbon to a large extent.

In carrying out his furnace operations, Mr. Howden endeavoured to approximate as nearly as possible to these conditions, and to work the furnace as, what may be termed, a combined quick-gasifying and complete-combustion furnace, by the following means:—The air in the ash-pit, with a given area of air-space through the fire-bars, and a given average depth of fuel, is maintained at a pressure designed to pass a quantity of air through the fuel sufficient to gasify it and bring it to the surface largely in the form of carbonic oxide. The air in the casing between the two furnace doors is maintained at a considerably higher pressure than in the ash-pit, and is thus received by the distributing boxes inside the furnace plate and inner furnace door. The air then, at a considerable temperature and at a high velocity, issues in minute streams from small holes in the interior side of the air-boxes, their aggregate area being proportioned to the normal work of the furnace, and their position arranged to cause the air to strike the fuel with force equally over the surface within the limits of the firebars, as represented in the illustration.

By means of these differential pressures, used as described, the weight of air required for the complete combustion of a given weight of fuel can be made much less than is necessary in an ordinary furnace, while the complete stage of combustion being chiefly on, or above, the surface of the fuel, a clear white flame and intense heat is generated where most effective for radiation, and most innocuous in its effect on the furnace-bars.

The grave disadvantages which must attend the use of the closed or air-pressure stokehold system, used so extensively in our war-ships, have been already pointed out, and it is a singular circumstance that though this system has been fitted in a number of vessels for several years, there is no record showing that for forced draught rates of combustion it has ever been tested by twenty-four hours' continuous working at sea.

The case of the steamers belonging to **Mr. Alfred Holt,**

which are fitted with closed stokeholds, and which have been worked more or less at sea, is not a case in point, for these steamers are not worked at a forced draught rate of combustion.

Mr. Holt has furnished particulars of two of these steamers, the *Hector* and the *Anchises*. The *Hector* develops 780 IHP from 90 square feet of grate in a boiler having six furnaces and 2504 square feet of heating surface, and the *Anchises* 900 IHP from 90 square feet of grate in a boiler with six furnaces and 3101 square feet of heating surface, being respectively 8·66 and 10 IHP per square foot of grate per hour. With this rate of combustion the injurious effects arising from this system of forced combustion are not developed, and the boilers are also, from their design, peculiarly fitted to withstand the injurious effects of the cold air admission.

The Holyhead mail steamers are also fitted with this system. Their boilers are quite capable of supplying full steam by natural draught, and the supply of air by the fans is only used when the weather is unfavourable or other circumstances prevent them from obtaining a good natural draught. The run each way occupies about three and a half hours, and the steamers are laid up one week out of every three. The rate of combustion gives about 6·75 IHP per square foot of grate, but this is with low-pressure simple engines.

These examples and similar ones in America, where working a fan in an open stokehold has been long in use, do not therefore fall within the cases of forced draught, so that the question whether boilers can be worked with this system, when the steamers are put to the use for which they were built, remains yet to be proved. Meantime, in the absence of direct proof, the effects of this system can only be judged of from experience obtained from analogous cases, and the experience of such cases leads to the conclusion that such a system is unfitted for the ordinary working of boilers at sea.

The cause of injury to boilers, common to both the closed stokehold system and to the cases recorded, is the sudden cooling of the interior of the boiler by a rush of cold air immediately after being under a high furnace temperature; but, in those boilers in which the ruinous effects of this sudden change of temperature have occurred, being worked by natural draught, the rush of cold air through the boiler was much less in velocity and volume than obtains in boilers with closed stokeholds under air-pressure.

The boilers referred to were fitted with their engines by Mr. Howden's firm a few years ago in steamers of over 4000 tons, whose names it is not necessary to give here. The boilers in each steamer were ten in number, made to the owning company's specification on a certain patent design somewhat of a locomotive type, but without sides or front to the fire-box part, which was formed of brick-work. The maximum working pressure was 125 lbs., and the boilers were designed to supply steam to work the engines up to 3300 IHP.

On trial, with the boilers quite clean and containing fresh water, they supplied steam steadily and well, and worked continuously over the whole day without the slightest leakage at any part or sign of injury. Starting thus on a voyage to Calcutta, the power being reduced to about 2100 IHP, the boilers ran for a number of days without trouble; but as soon as the formation of a very slight scale on the tube plates began, and the water increased in density by replenishing the waste from the sea, the increase of temperature on the tube plates exposed to the flame rendered these so much more sensitive that, on the opening of a furnace door, the sudden rush of air across the furnace, a distance of 6 feet 6 inches, struck the tube plate, with a difference in temperature sufficiently great to cause a sudden contraction of the tube plate and unequal contractions of this thicker tube plate and the thinner tube ends, so that a general leakage at the tubes took place. This leakage once begun could not by any means be stopped, so that the dust and soot from the fire adhered to the damp plate, and in a short time accumulated to such an extent as almost to close up the tubes and reduce the power of the boilers to a small fraction. The only course then left was to draw the fires, clear and tighten the tubes with an expander, and then go on again to repeat the above process, sooner or later, even with the most careful management. To work these boilers, without continual trouble and great expense, even at a further reduced power, it was found necessary to use fresh water only and to protect the tube plates as far as possible by loopholed brick-work. A large multitubular boiler, having two furnaces 3 feet in diameter, was kept employed in supplying steam from sea-water for condensation in order to make up the entire waste from fresh water. Notwithstanding these precautions, and after many expensive repairs, the boilers of the first of these steamers were, after eighteen months' working, so much injured that new boilers of ordinary form were ordered to replace them. Before their completion this steamer was lost in a voyage across the Atlantic, and the new boilers were fitted into the second steamer a few months thereafter, the boilers of this latter vessel having suffered in a like manner to that described.

These boilers, though of a type more liable to injury than the usual marine type, show what would undoubtedly take place in any boiler in which a large volume of comparatively cold air suddenly impinged on the interior plates, especially the tube plates, which, until the moment of impact, had been under a high furnace temperature. That the vast volume of air rushing through the furnace of a boiler worked on the closed stokehold system, with forced combustion, would strike the tube plates cold enough to cause serious damage, may be inferred from the area of the volume and its velocity—the former that of the furnace door, the latter 60 to 70 feet per second.

This conclusion is indeed not left to mere inference—it is a fact within the experience of most sea-going engineers working with chimney draught, that when considerable scale accumulates on a back tube plate of an ordinary marine boiler, if a furnace door is suddenly opened under certain conditions of draught, the rush of cold air causes the tubes to leak. It will be remembered that the boilers just described, which suffered so greatly from the effect of the impact of cold air, did not show the slightest injury on a prolonged trial nor for some days after they had been at sea, while in the *Satellite* and *Conqueror* a few hours' trial, under the most favourable conditions, produced the injurious effects described.

No doubt it is stated that it is not intended to work

the steamers with this forced draught at sea unless emergencies arise to require it, because the boilers are made large enough to supply steam for full power by natural draught. If so, then the proper advantages of forced combustion, reduction in weight of machinery and space occupied, are given up, and the vessel cumbered with an expensive and troublesome arrangement, well known to be at least wasteful in fuel, for the prospective advantage of it being used only occasionally and in case of emergency. But will even this prospective advantage be realised? It will not; for after a steamer fitted on this system has been at sea for some time with natural draught, and the boilers have accumulated a slight scale, if it were attempted to use forced draught on an emergency, the boilers would soon be rendered useless, and the vessel left helpless.

Another difficulty which is never experienced on trials of a few hours, or on short sea passages, but which must be faced at sea with continuous working, is that of cleaning fires. With the large consumption required for this system of forced draught, fires could probably not run for more than six hours without cleaning. Would it be possible to clean a fire with a volume of cold air of the area of the furnace door rushing through the furnace and tubes at the velocity due to $1\frac{1}{2}$ inches or 1 inch water-pressure? Supposing the boiler was not injured by the terrible scour of cold air, it would certainly be cooled down to a serious extent by the furnace and tubes acting, for the time being, as condensers. The only alternative would be to reduce the air-pressure, for the time being, to that of the atmosphere, if this could be done with safety in a boiler-room in connection with others through a common funnel. Take a steamer having thirty-six furnaces divided into four separate air-pressure compartments, cleaning fires as mentioned, six furnaces would require to be cleaned every hour. This would place at least half of the boilers continuously under very low combustion; so that whatever alternative, whether this or the worse, was adopted, the cleaning of the furnaces on this system, under continuous working, could not be other than a serious practical difficulty.

Putting aside altogether the question of injury to the boilers, and recurring to the fact that the steamers of the navy which are fitted with this forced draught system have boilers of the usual size, so as to steam full power by chimney draught, let us assume that with forced draught the boilers of a steamer of the larger class could supply steam to drive the engines continuously at sea at a power 20 per cent. above that of full chimney draught power, without trouble in cleaning fires or the slightest injury to the boilers. Even in such a case as here assumed, the same steamer fitted on Howden's system of forced draught, and with the boilers required for equal power, would steam at equal speed three times the distance it could do with the closed stokehold system of forced draught. This would be owing, first, to the greatly reduced size and weight of boilers, water, and fittings required with his system, leaving room for larger carrying space; and, secondly, to the much smaller consumption of coal required to produce the same power.

The vital importance of such reduction of weight and space and coal consumption in war-ships does not require illustration here, and it cannot be disputed that the difficulty in building ironclads on favourable forms for economical propulsion, and securing at the same time sufficient protection in thickness of armour plate, would be greatly obviated by the reduction of 1000 tons in the weight of machinery and coal required in one of the larger ironclads. Still keeping out of view the vital difference between using a safe system adapted for everyday work, and an unsafe system unsuited for such work, the advantages of possessing high-speed cruisers, able to run continuously with ease at the highest speed and to keep the seas, for thrice the time they could with the present system, may be safely left to be estimated by naval officers and others experienced in handling such ships.

For steam yachts the use of Howden's system of combustion would make a speed possible, and with a reduced weight of machinery, that is quite unattainable with present arrangements, while the absence of smoke, the perfect control of the steam—so that blowing off at the safety valves under any circumstances can be prevented—the power to reduce the evaporation at will to the lowest point sufficient to turn the engines, and to lie at rest for many hours with the combustion practically stopped, and thereafter to go on at full speed at one minute's notice, are all advantages of a most important character, not only for yachts, but for any class of steamship, and more especially for ships of war.

Indeed it may be said that the use of this original system of forced draught is now being freely adopted wherever sea-going steamers are meant to successfully and profitably maintain their position in the severe competition of the present day. For the number and size of boilers required by this system, for equal power, being only from two-thirds to one-half that required for natural draught, the great reduction in space required for boilers, and the proportionately reduced dead weight in boilers, water, and fuel carried, make it a great desideratum for the highest-powered ocean steamers. When such steamers adopt this system, they can carry considerably more passengers and cargo, and at much less expense, so that they may be run at a profit on rates which would be found absolutely unremunerative for similar steamers with natural draught only.

Highly economical results have been obtained in boilers with this system on two square feet of heating surface per IHP, and at rates of combustion giving from 20 to 24 IHP per square foot of fire-grate, with grates up to 5 feet 6 inches in length.

The Howden system of forced draught unites correct principles of combustion with effective working arrangements, producing, when properly worked, complete combustion with a reduced weight of air, besides utilising the waste heat. The closed stokehold system, which is almost exclusively confined to war steamers, is both injurious and wasteful, and is now being almost universally discontinued.

When Mr. Howden read his first paper on "Forced Draught" at the Institution of Naval Architects in 1884, describing his system and giving the general results of two years' experimental working in marine boilers, there was not one sea-going steamer in the United Kingdom or in any other country fitted or being fitted with forced draught. The particulars given in that paper, and the highly successful and economical results of the West India steamer, *New York City*, which began running shortly after, and of the steamers which followed, first called attention to the importance of forced draught for mercantile steamers. That its advantages have now become widely recognised is sufficiently shown by the number of steamers to which it has been applied for. Besides, a large number of vessels owned by private firms, in some of the largest companies in the United Kingdom, are having it fitted to a considerable number of their steamers.

CHAPTER VII.

Liquid Fuel — Its Early Introductions — Geographical Distribution — Theories as to its Origin — Chemical Constitution — Thermic Value compared with that of Coal — Possible Efficiency — Cost of Oil to admit of its Economical Use — The Value of Increased Cargo Space — Dangers of Stowage — Precautions to be observed in Oil Storage — Methods of Application — Injectors — Use of Oil under Pressure — Forced Draught Oil Firing System — Tabulated Result of Trials — Advantages and Disadvantages.

A CAREFUL analysis of the character of the present tendencies toward economy in the methods of generating motive-power appears to show that, in *all* fields of application, the movement is conspicuously in the direction of concentration of energy.

With any kind of fuel, the intensity limit is, of course, contracted by the sum of its heat-producing constituents; with liquid fuel we have a great and *primâ facie* advantage, in the higher thermic value of its constitutional heat-producing elements—the hydro-carbons of a high molecular density forming a great percentage of this mobile and wonderful fuel.

That this liquid or mobile character of fuel will eventually, and in the near future, be in one form or other the motive-power creator for marine propulsion purposes may be accepted as an accurate forecast.

Premature and ill-conceived attempts have from time to time been made to practically introduce the application of liquid fuel for marine purposes, but the comparatively limited supply then available soon had such an inflating effect on the cost of the fuel as to absolutely prohibit its use on economic grounds.

The obstacle of scarcity of supply is now rapidly falling away under the influence of the almost universal discovery of this wonderful natural fuel, and the phenomenal increase of production during the last few years.

To indicate the widespread occurrence of this natural fuel, one has merely to mention the geographical *locale* of actually proved oil fields. In Europe there is Roumania and Galicia; in Asia Minor, the famous Caucasian region of fire; in Asia, Persia, Burmah, India, Beloochistan, Japan, and China; in Australasia, New Zealand; in South America, Venezuela, and fields of promising copiousness in Peru; in Africa—both North and South—oil has been found, and in North America—both in the United States and Canada—there are indications of an almost exhaustless and colossal supply.

According to Mr. B. H. Thwaite, it appears that the income derivable from these natural liquid fuel resources enabled the Northern States of America to recover very speedily its financial prestige after the terrible drain of the Civil War, and the expression "struck ile" is now synonymous of great good fortune. There is no doubt but that the bountiful gift of nature to Russia on the Caucasus has staved off the national bankruptcy of that country.

There are several theories put forth accounting for the origin of this liquid form of fuel. They may be divided into two categories—one identifying the origin with the metamorphosis of vegetable and animal marine organic matter, the other attributing the origin to synthetic and mineral effects—as distinct from those of an organic genesis. Mr. Thwaite, after an analysis of the different theories, suggests that, as coal is the result of inland animal and vegetable organic matter, therefore petroleum oil is the result of prolific animal and vegetable submarine organic life, which, under the influence of physical compression and interrestrial heat, is partially distilled, the higher olefiants of the hydro-carbon series forming the basis of the natural gas, which is nearly, if not always, associated with the oil and strata. The heavier hydrocarbons remain in solution, probably a state induced by the chemical influence of the saturated solution of chloride of sodium or salt water. He has also drawn attention to the geologic fact that all the great oil and natural gas fields are associated with salt or brine fields.

In China, from almost prehistoric ages, the salt brine has been evaporated by natural gas conveyed by pipes from below the salt strata.

Petroleum oil is a hydro-carbon, and the varying characters of oils depend for their different values upon the ratio of the hydrogen to the carbon. A sample of the crude Russian astatki was carefully analysed, the average constitution giving—

Carbon,	84·940 %
Hydrogen,	13·960 ,,
Oxygen, by difference,	1·255 ,,
	100·155 ,,

The formula for oil may be taken to have a ratio represented by the symbol HC_7 (that is, one volume of hydrogen to seven of carbon).

Triple and Quadruple Expansion Engines

For practical purposes, the value of crude petroleum, thermically expressed, may be taken at 20,000 BTUs in comparison with best Nixon's coal taken at 14,000 BTUs; then the ratio may be taken as 7 is to 10 in favour of oil. There is, however, another advantage represented by the higher factor of efficiency attainable by the use of oil fuel. Generally $\frac{4}{7}$ of the heating power of coal, or 57 per cent., can only be attained in practice, whereas with oil fuel $\frac{7}{10}$ to $\frac{8}{10}$, or 70 to 80 per cent., of the full thermic efficiency of the oil has been obtained.

The relative values of the two fuels, from an economic measurement, is given in a contribution entitled "Crude Liquid Hydro-carbons for Heating and Lighting Agents." In this contribution, for which we are indebted to Mr. Thwaite, the coal is estimated to cost 12s. 6d. per ton delivered. Firemen's wages are taken at 1s. per ton fed into furnaces. The chemical constitution of the coal is taken to be represented by the following:—

Carbon,	80 %
Hydrogen,	5 ,,
Oxygen,	8 ,,
Ash, by difference,	7 ,,
	100 ,,

The calorimetric value of the coal is taken at 14,000 BTUs. The petroleum is of the crude variety, satisfying the equation already given, and having a calorimetric value of 20,000 BTUs.

In the following table the relative higher efficiency of the oil is not allowed for. But there are many advantages to be enumerated that should be credited to the oil,—for instance, the greater efficiency attainable in practice, and the reduced cargo space required for fuel. The plus signs represent comparative advantages of use, and the minus signs the disadvantages. The cost of supervision of the oil-firing apparatus is taken at 2d. per ton.

Cost of Oil. Per gallon in pence.	Cost of Oil. Per ton in shillings.	Use of Oil. Percentage of advantage or *vice versâ*.	Use of Coal. Percentage of advantage or *vice versâ*.
s. d. 0 0½ 0 0¾ 0 1 0 1½ 0 2	s. d. 10 0 15 0 20 0 30 0 40 0	+ 40 % + 36 „ + 15 „ − 28 „ − 70 „	− 40 % − 36 „ − 15 „ + 28 „ + 70 „

It appears, therefore, that the neutral line falls between the price of oil at 20s. per ton, or 1d. per gallon, and 30s. per ton, or 1½d. per gallon. It may be safely assumed that wherever crude petroleum is obtainable at 25s. per ton it is economical to use it, in comparison with coal at 12s. 6d. per ton delivered f.o.b. Allowing, however, for the increased efficiency attainable with the use of oil, it would be economical to employ petroleum oil at 30s. 9d. per ton, in comparison with the cost of coal at the figure stated, of 12s. 6d. per ton f.o.b. The density of the oil varies according to the density of the hydro-carbons of which it is composed, and its specific gravity may be taken to range from ·88 to ·94.

	Chemical Constituent of Oil.
Light, ...	$\begin{cases} C. = 86·3 \\ H. = 13·6 \end{cases}$
Heavy, . . .	$\begin{cases} C. = 86·6 \\ H. = 12·3 \end{cases}$

156 *Triple and Quadruple Expansion Engines*

We may assume that 1 ton of petroleum oil requires a stowage space of 39 cubic feet, whereas the navy allowance for bunker space for coal is 48 cubic feet to the ton.

On the stowage of 1000 tons of fuel the cubic space saved, and available for cargo by the use of petroleum oil, can thus be calculated—

$$\frac{1000 \times 39 \times 7}{10} = 27300$$

and $27300 - \left(\frac{23 \times 27300}{100}\right) = 21021.$

Then $\left(\overset{\text{Coal.}}{1000 \times 48}\right) - \left(\overset{\text{Oil.}}{21021}\right) = 27980$ the cubic stowage space saved.

For long distance ocean voyages this great saving in storage space is of substantial advantage, either in enabling a larger quantity of heating power to be stored, or in permitting a greater tonnage of cargo to be carried. Related to this advantage there is another, which is that the oil can be pumped into the tanks, thereby avoiding the expensive and often troublesome labour of trimming coal.

There are, however, dangers attending the storage of oil if not properly arranged, and these originate from certain physical and chemical characteristics of the oil. The refuse petroleum, from which the more volatile hydro-carbons have been expelled, has a very high flashing point, and does not volatilise below 400° Fahr. The crude oil from the wells, however, volatilises at about 140° Fahr., and some of the oils flash at 70°, consequently the storage tanks should be properly constructed, so as to allow the gas to safely escape, but without permitting any air to mix with it, or to afterwards enter the tank.

According to St. Claire-Deville, the co-efficient of

expansion of oil is 0·0006 to 0·0009 per degree centigrade of increase of temperature, so that 1000 cubic feet of oil raised, say 30°, increases in bulk by 24 cubic feet, so that it is necessary to allow 5 per cent. additional tank or trunk space for expansion. Oil tanks should be self-contained, and be approachable on all sides for examination and repairs.

For land storage the Thwaite Safety Oil Tank removes all objections on the score of danger. In this tank an annular water cavity surrounds the oil. Over the tank there floats a balanced cover, which ascends or descends according to the expansion or contraction of the oil or the oil gases evolved, and prevents air entering and producing an explosive mixture.

There is no doubt but that recent explosions in oil-tank steamers have been produced by the introduction of air into the tank, which, becoming mixed with oil vapour, formed an explosive mixture: such a contingency, therefore, as the introduction of air should be made impossible, and oil storage would then be rendered almost as safe as the storage of coal.

In proceeding to describe the method of application of oil or liquid fuel to steam-raising purposes, we may divide them into four broad classes:—

1. The gravity method, by which the oil is allowed to flow on an inclined serrated trough, or fall from an elevation in their sheaves, around or over which air is drawn by the natural chimney draught.

2. The pulverisation method, in which the combustion of the oil is effected by the spraying of the fluid by the aspiration or introduction of steam or compressed air; special apparatus, known as oil injectors, are employed

for this purpose, the principal types known being Urquhart's, Henwood's, Holden's, and Thwaite's.

3. The conversion of the oil into a gaseous condition before allowing it to meet the air previous to ignition.

4. The method of forcing the oil in small streams through small orifices, by means of simple plunger force-pumps, and projecting such stream on to fire-brick blocks in such a way as to produce spray, which is then caught by a current of air, combustion being then completely effected.

It will be readily observed that the methods classified under heads 1 and 3 are impracticable and inappropriate for marine purposes, except for coasting steamers. Method 2, when involving the use of steam which cannot be condensed, has disadvantages, which, however, are removed by the use of compressed air. For coasting steamers, the steam jet oil injector is an exceedingly handy contrivance, its simplicity being all that can be desired, and the more effective as sprayers the injector is the better. The latest type, known as Thwaite's Rifled Injector, produces a circulatory flow of the spray which, by centrifugal action, flows from the nose of injector in a widening circle, and permits the air to be intimately associated with the oil. This is important, because the high hydro-carbon density of the oil has been found to present a difficulty in securing perfect combustion. Mr. Holden's injector is provided with an auxiliary steam ring fixed around the nose of the injector, and this ring of steam assists in comminuting the oil, and gives a greater control over the intensity of combustion; the weight of steam used would, however, be a serious objection in the application of this injector to marine purposes.

Lately, Mr. Thwaite has successfully developed the method identified in Class 4. He employs a simple 3-throw force-pump, in the outlet of which there is interposed a weighted valve, which insures a pressure of 80 lbs. to the square inch on the oil. This oil, under pressure, is led into a rifled injector, into which a small volume of air is compressed (or steam may be employed), and this assists in spraying the oil jet, and, projecting it on to a piece of fire-brick, which effectually completed the work of disintegration. This method has the advantage of enabling all the steam to be recondensed, so that the great difficulty of using oil for long voyages is entirely overcome.

An application of the system of the Gaseous and Liquid Fuel Supply Company (of Manchester) to a torpedo boat is shown in Fig. 47. This figure is a good illustration of the most modern methods of applying liquid fuel. It will be seen that the air is supplied under pressure, two "Blackman" fans being used for this purpose. The air is delivered to the hollow parts of the fire-door frame, and becomes highly heated; it flows into cone-shaped chambers, in which the oil injectors are fixed; the heated air thus flows round the oil spray, and the combustion is highly perfect, the line of spray is directed on to a bed of coke and broken pieces of fire-clay placed on the grate; this latter enables the steam to be raised, in the first instance, with wood or coal, or other waste, dipped in oil. The injectors are adapted to be fed either with steam or with compressed air.

There is a heat accumulator chamber of fire-brick located at the end of the fire-grate; this is maintained at a temperature of white incandescence, and served not

160 *Triple and Quadruple Expansion Engines*

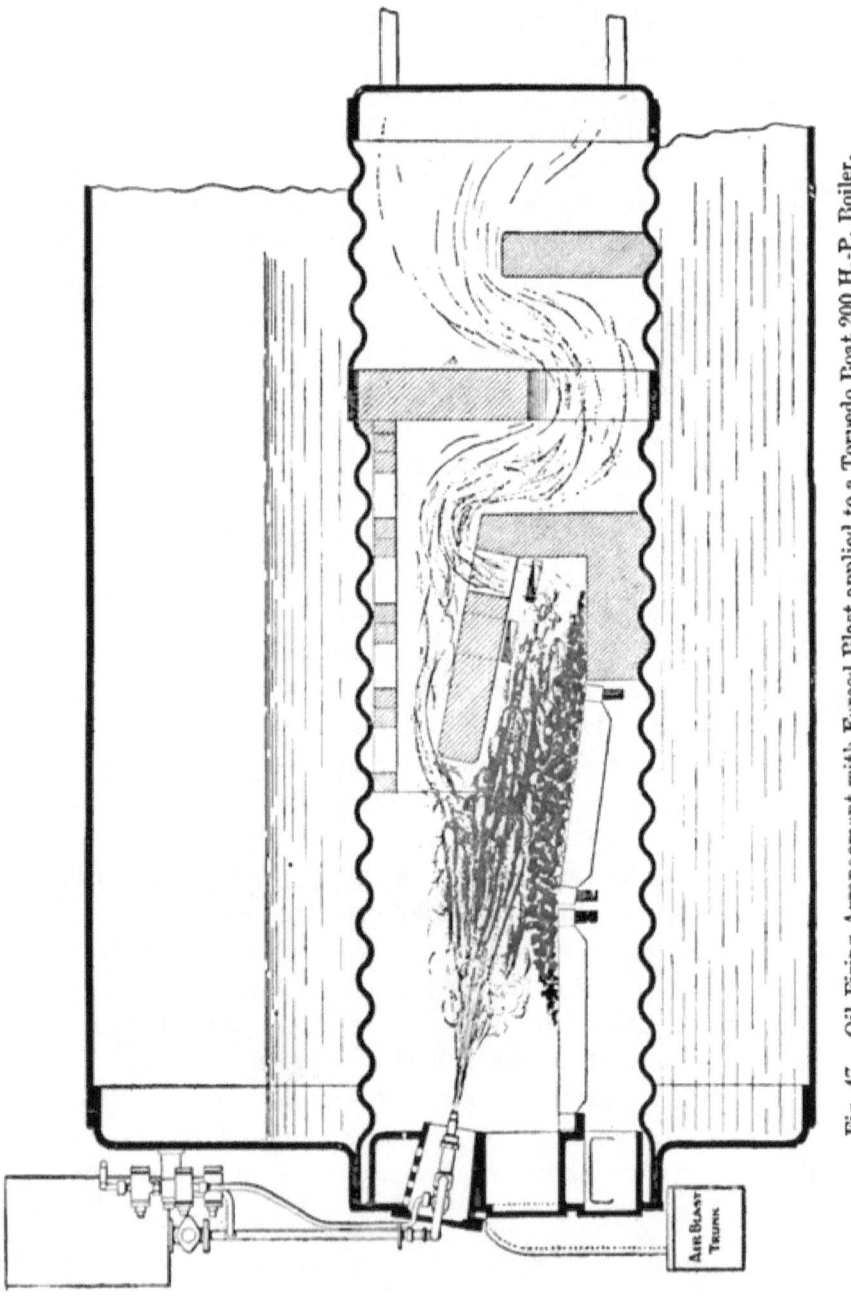

Fig. 47.—Oil Firing Arrangement with Forced Blast applied to a Torpedo Boat 200 H.-P. Boiler.

only to equalise the rate of steam production, but prevents the passage of the unburnt or unoxidised hydrocarbons into the tubes of the boiler.

Valves are provided to regulate the oil, steam, compressed and forced draft air supplies, so that the entire action is under complete control, and in regular running the safety-gauge needle rarely moves.

The following table, for which we are indebted to Mr. B. H. Thwaite, gives the results of the two days' experiments in oil firing of a return tube marine boiler.

The calorimetric value of the oil used was equivalent to the evaporation of 16·66 lbs. of water per lb. of liquid fuel, or 1 to 16·66 from and at 212° Fahr.

N.B.—Figures in brackets denote the lbs. of water actually evaporated from and at 212° Fahr.

	First Day's Experiment.	Second Day's Experiment.
Net actual evaporative efficiency . . . =	[14·97] 89·87 %	[14·21] 85·25 %
Net useful evaporative efficiency after deducting steam for aspirating oil =	[14·33] 85·99 %	[13·58] 81·50 %
Net thermic efficiency including the heat carried away by the products of combustion, and that necessary to elevate them up the chimney . =	[15·90] 95·42 %	[15·42] 92·60 %
Heating surface efficiency in lbs. of water evaporated per sq. foot per hour =	2·392	1·936
Heat absorption efficiency tested by Carnot's law when Average initial temperature of combustion flue =	1720° Fahr.	1700° Fahr.
Average final chimney temperature . . =	248° Fahr.	262° Fahr.
Then	$\frac{1720-248\times100}{1720}=85\cdot57\%$	$\frac{1700-262\times100}{1700}=84\cdot53\%$

In his address to the Liverpool Marine Engineers Institute, Mr. Thwaite pointed out the desirability that oil tanks should be erected at the principal coaling stations, along with an adequate piping and pumping equipment.

The tendency of the oil market is a continual lowering one; at Baku, the supply is exceeding the demand, and it may be safely asserted that in the next few years the price of oil will recede to such a figure as to enable its economic use on the Mediterranean, Indian Ocean, and Australian fleets. When it is realised that the value of the crude petroleum at the wells is only one copeck per pood, or about $\frac{1}{2}$d. per ton, it will be easily realised, that at the source of the output, it has an enormous advantage over coal, which requires getting, whereas in most of the wells the oil flows like water from a fountain, and has merely to be tanked and piped to the first railway or shipping depôt, or to the refinery. Therefore, the time may not be far distant when the cost of oil will enable its use to be economically effected for marine purposes.

We may now summarise the advantages and disadvantages of liquid fuel for marine steam boiler firing:—

The advantages are:—1. Reduction in weight of fuel for a given heating power. 2. Reduction in absorption of valuable cargo space. 3. Greater efficiency of steam generating surface. 4. Less wear and tear of steam boiler, reduced expansion and contraction of flue plates, because the doors do not require to be opened. 5. Less physical effort, greater independence of labour. 6. Greater control over combustion effect. 7. Uniformity of rate of steam production. 8. Greater cleanliness, no ashes or clinkers to clear away. 9. Absolute smokeless combustion can be obtained. 10. Fires can be extinguished at a moment's notice. 11. Far easier supervision. 12. Oil can be used in cases of extreme urgency for sea-calming purposes.

The disadvantages may be stated as :—1. Greater care required in storage. 2. More intense combustion, and greater strain on crown of furnace. *N.B.*—This can be overcome by the use of the Saddle Circulator (Thwaite's patent). 3. Greater present first cost at the ports of the U.K. 4. Limited number of oil storage installations at present available for marine purposes. 5. Uncertainty of price.

By the method known as Henwood's system, it is claimed that perfect combustion and great economy result from the plan of vaporising the oil-fuel by means of the double jacket of superheated steam just prior to its entering the fire-brick chamber in the furnace tube, and on the vaporised oil passing out of the injector it becomes so effectively diffused by and with the superheated steam, the component elements of which, when burning, augment the evaporative power of the hydrocarbon oil-fuel.

By dispensing with the fire-bars the objectionable cold ash-pit is abolished, and the whole of the furnace-tube is available as heating surface; while the properly designed fire-brick chamber so evenly distributes the heat evolved that no injury is caused to the furnace-tube or the tube ends.

By this system it is possible to keep the furnace door closed when once the furnace or oil-fuel gases have been lighted, and to dispense with labour.

There is no valid objection to the employment of oil-fuel in a vaporised state for marine or other purposes, as by its employment in a vaporised condition the greatest economy can be effected in conjunction with the superheated steam, the dissociation of which has been found to largely augment the evaporative power of the hydrocarbon oil-fuel.

The objection made to the use of steam for such purpose on ocean-going steam vessels is obviated by the employment of efficient evaporators, which can supply sufficient condensed water at trifling cost.

As to the employment of compressed air, it is known that air, being composed of about one-fifth part of oxygen and four-fifth parts of nitrogen—which latter neither supports life nor combustion—is extremely detrimental to the efficiency of the oil-fuel and absorbs heat, and so reduces the heat available for evaporation.

The late Professor W. J. Macquorn Rankine, who contributed some valuable papers on this subject, gives as the theoretical evaporative value of burning hydrogen 64 lbs. of water evaporated against 20 lbs. by petroleum, or 15 lbs. by coal.

By the most efficient system an evaporation exceeding by over four times that obtainable from best Welsh coal (on the average about 8 lbs. on a voyage) has been obtained.

CHAPTER VIII.

The Marine Boiler—Quantity of Steam Generated and Fuel Consumed—Perfect Combustion—Size of Fire-Bars—Length of Fire-Grate—Height of Bridges—Circulation in a Boiler—Prevention of Scale—Constituents of Scale—Evil Effects of Scale—Collapsed Furnaces—Corrosion and Pitting—How Caused—Chemical Composition of Corroded Parts—Methods of preventing Corrosion—Hannay's Patent Electrogen.

IT is now generally admitted that the most important duty of a marine engineer is to bestow every care and attention upon the boiler. The boiler proper consists essentially of two parts—there is first the furnace, with its combustion chamber, in which the fuel is consumed and heat generated; and then there is the space in which the water is heated and thereby turned into steam. The quantity of steam generated depends, in the first place, upon the quality of the fuel and the quantity burnt. The quantity of fuel consumed depends upon the area of the fire-grate and the quantity of air admitted into the furnace. The efficiency of a boiler depends upon securing absolute contact between every portion of the products of combustion and the plates, and leaving nothing, or next to nothing, to be done by radiation. The efficiency of the furnace depends on its power of burning, with as little waste as possible, the whole of the fuel supplied to it, and to produce perfect combustion there must be as little waste of heat as possible in obtaining the necessary draught. To insure this the fire-bars should not be crowded as thickly into the furnace as they will go, but there should be sufficient space left between them to allow admission of the quantity of oxygen required for complete combustion of the fuel. Fire-bars are generally made too thick and not deep enough—they ought to be only about $\frac{5}{8}$ inch thick at top, tapered away to $\frac{1}{8}$ inch at bottom, and not less than 5 inches in depth. The fire-grate is also generally too long, and where the fires do not require to be forced it will be found that in many cases it may be shortened with great advantage. A long grate will raise more steam in a given time than a short one, but the increased evaporation will not be equal to the increased consumption of coal.

The bridges should not be made flat but arched at the top, as this insures a better distribution of the gases on the furnace crowns and in the combustion chamber. If a bridge is too low there is a waste of fuel through the gases not coming in contact with the furnace crown, and if too high the velocity of gases may be increased so much that they will go up the funnel, without doing any useful work through taking with them the greater portion of their heat. A fairly good proportion is to make the area of opening over the bridge about equal to one-sixth the area of the fire-grate; but experience will best decide the most efficient area for each particular boiler.

It has been proved by experience that a furnace may consume a large quantity of fuel without perfect combustion taking place, and when it does take place only a comparatively small portion of it may be usefully employed. One lb. of coal can be made to evaporate 14 to 15 lbs. of water, but in practice 10 lbs. is seldom exceeded, showing that even when at its best the efficiency of the boiler is comparatively small; but if there is either an excessive or insufficient supply of air admitted, or a bad draught or bad stoking, the efficiency will be still further reduced. When combustion has been completed in the combustion chamber the evaporative power of that portion of the boiler is high, and it may be reckoned the most efficient for transmitting heat to the water.

The internal efficiency of the boiler depends, to a great extent, on the circulation of the water to and from the heating surfaces, it being a well-known fact that a boiler with a defective circulation is not nearly so efficient in raising steam as one in which the circulation is good.

It should also be borne in mind that a clean boiler will generate steam much more rapidly, and with less fuel, than one which is covered with scale. Keeping this in view, and remembering that with the high pressures now in use scale deposits much faster, becomes much harder, and is much more dangerous on account of the extremely high temperature, the problem has now become, how to prevent scale from forming, not how to remove it after it has been formed. There are two methods in ordinary use for preventing the formation of scale, which we may describe. The first method was introduced by Professor Lewes, who, in 1889, read a paper at the Institution of Naval Architects which aroused considerable interest at the time, and in which he traced the formation of the various kinds of boiler deposit, and pointed out that it was possible to so prepare sea-water that it should give no incrustation unless evaporated to a density at which the salt itself would crystallise out. Having devised an apparatus for this purpose, Mr. J. H. Biles, the general manager of the Naval Construction Works, Southampton, determined to try the process in a boiler working a crane on one of the jetties in his yard, in order to practically determine if the method would give the results claimed for it.

The boiler was kept under steam for a month, using nothing but prepared sea-water, and blowing off the steam when it was not required for other purposes. It was then opened in the presence of a representative gathering of gentlemen interested in the subject, and the result was found to be in every way eminently satisfactory, the interior of the boiler being in perfect order and free from any trace of incrustation or scale, the plates merely looking as though they had received the thinnest possible brush over of whitewash.

The water used in this experiment was drawn from the Itchen a short period before high tide, enough being stored in a large tank to last for the day's consumption, and on analysis it proved to be practically sea-water contaminated with sewage, and containing a sufficiently large proportion of suspended earthy matter, to have given, under ordinary circumstances, an extremely heavy incrustation. The water was pumped a ton at a time into the precipitating vessel, which consisted of an egg-shaped boiler placed on end, with a small man-hole at the top for the introduction of the precipitating materials, and fitted below with blow-off cock, a tube to lead away prepared water, and a pipe through which steam could be blown into the contents of the boiler, or "precipitator," as it is called.

The water having been introduced, exhaust steam was blown into it until the boiling point was reached, and the "precipitator powder," consisting chiefly of carbonate of soda, made up in packets of the right size, was then introduced through the small man-hole, and entering into solution in the water, it immediately threw down all the lime and magnesium salts present in the form of a white flocculent precipitate. The action, however, was not yet quite completed, the precipitate not being dense enough to settle rapidly; these points were, however, attained by closing the man-hole and blowing in steam until a pressure of 10 lbs. was obtained in the precipitating vessel, and this pressure being maintained for a short period the precipitate became very dense and rapidly settled, merely leaving a slight turbidity, which was got rid of by running the prepared water through the asbestos cloth in a filter-box. Water so treated has a density of $\frac{1}{32}$, contains nothing but sodium salts, and may with safety be evaporated until its density reaches $\frac{5}{32}$. These salts, being excessively soluble, will not form any deposit until sufficiently saturated for the salt itself to crystallise out, the salts present not crystallising out until a density of from $\frac{6}{32}$ to $\frac{7}{32}$ is reached.

The advantages of such a prepared water for use in marine boilers are, in the first place, cheapness, as the preparation only costs 1s. 2d. a ton, or less than is paid at many ports for ordinary fresh water, and far less than distilled water can be made at by any process. Secondly, it saves the carriage of either considerable quantities of fresh water or else of complicated distilling plant, whilst it is preferable to distilled water, the solvent action of which upon metals is a distinct drawback. Thirdly, it not only does away with all incrustation, but, by also removing the magnesium salts present in the water, prevents pitting, and does away with the necessity for zinc in the boiler; and, finally, the density of the prepared water being less than that of sea-water, and it being possible to evaporate off a much larger proportion of it—the loss of heat, by blowing off, is very small; and, it may be added, that as the exhaust steam used for heating the sea-water is all condensed and passes back with the prepared water, which is at the boiling point, to the hot well, loss of heat is also here avoided.

The second method, and the one in most general use, for preventing the formation of scale on the heating surfaces, is to use distilled water. As already stated, it was found, upon the introduction of high-pressure boilers that the quantity of scale deposited was so serious, and the difficulty of removing it from between the tubes and narrow water spaces so great, that it became almost absolutely necessary to incur the expense of fitting up apparatus for supplying the boilers with distilled water rather than run the risk of burnt plates and collapsed furnaces. After some years' experience of this system it was found that the trouble had only been transferred from the boilers to the distilling apparatus, and that the enormous deposits formed in the latter frequently caused them to break down or become in some way unfit for use, thus necessitating the introduction of sea-water into the boilers to eke out the supply of distilled water from the condenser.

Distilled water is presumably free from all impurities, but condenser water always contains some of the lubricants from the engine cylinders, and in the days of animal and vegetable lubricants, or fat and oils other than mineral, this water, on account of the superheated steam breaking the oils up and liberating fatty acids, caused a large amount of damage to the plates of the boiler; and even of late years, with the use of mineral oil lubricants, a new and serious trouble has arisen.

So far the deposits taken into consideration have been those formed from the impurities natural to the water itself; but Professor Lewes has pointed out that with the introduction of high-pressure steam a new and highly dangerous form of deposit was added to those already causing trouble to the marine engineer.

As early as 1878 the collapse of the furnaces of the s.s. *Ban Righ* and a similar misadventure in the screw tug *Ich Dien*, with no apparent cause to account for them, directed attention to the action which had taken place. The only clue found was a certain oily deposit on the tops of the furnaces, and experiments made by Mr. Dunlop of Port-Glasgow led to the conclusion that this oil, evidently distilled into the boiler from the lubricants used in the cylinder, was so bad a conductor of heat that its presence on the plates caused them to become superheated, with the result that, being unable to withstand the pressure of steam in the boiler, they collapsed.

After that date similar cases of collapse became quite frequent, no fewer than thirty vessels having been disabled from this cause during the last few years.

One case that came under notice was that of a large steamer trading between Liverpool and Boston, fitted with ordinary compound engines, and being twelve days on the voyage. She had three double-ended boilers, with three plain furnaces at each end, and three combustion chambers in each boiler. The furnaces were plain, in one length, and connected at the back end to the tube plates, being flanged up inside the chamber, whilst the front end plate was flanged inwards on to the furnace crowns. The furnace crown was $\frac{1}{2}$-inch plate, and the front bottom plate $\frac{11}{16}$ths, the working pressure being 80 lbs. The boilers were about five and a half years old, and had always been refilled with fresh water at the end of each run, both at Boston and Liverpool; whilst, as a rule, the waste on the voyage was made up by the use of about 70 tons of fresh water.

During the last voyage sea-water was used for this purpose, and every four hours whilst under steam 4 lbs. of soda crystals were put in the hot well, making about 2 cwt. during the run, the total capacity of the boilers being about 81 tons; while for lubricating purposes about 7 pints per day of valvoline were used in the cylinders.

When in port the boilers were allowed to cool down, and the water was run off, when they were swept down with stiff brushes, and afterwards sluiced out with a hose shortly before being refilled with fresh water. No trouble occurred with the boilers until five voyages before the final collapse, when some of the furnaces began to creep in; they were stiffened with rings and stays, but on the succeeding voyages the whole of the furnaces got out of shape, one after another. Examination of these boilers clearly showed that they had never been very heavily scaled, as in parts of the boiler where it would have been impossible to get at them to clear them out, no signs of heavy incrustation were to be found, and the absence of marks of scaling tools also showed that they had never been allowed to get very dirty. On the furnace crowns, where they had collapsed, there was only a slight white scale, not more than $\frac{1}{64}$th of an inch in thickness, whilst on the bottom of the furnaces there was a brown oily deposit $\frac{1}{16}$th of an inch in thickness, which in other parts of the boilers increased to between $\frac{1}{8}$th and $\frac{3}{16}$ths of an inch.

The boiler plates were as good as the day they were put in, and showed no structural signs of having undergone any change, whilst the analyses of the Liverpool water and the soda crystals used showed that they could have taken no part in the action which had led to collapse.

I. VALVOLINE.

The valvoline on analysis gave :—

Vegetable and Animal Oil,	nil
Mineral Oil,	100 per cent.
Acids (free),	nil
Boiling Point,	371° C.
Specific Gravity,	·889

II. SCALE FROM FURNACE.

	From Top.	From Below.
Calcic Sulphate,	84·87	59·11
Calcic Carbonate,	5·90	6·07
Magnesic Hydrate,	2·83	11·29
Iron, Alumina, and Silica,	2·37	2·85
Organic Matter and Oil,	3·23	19·54
Moisture,	0·80	1·14
Alkalies,	nil	nil
	100·00	100·00

III. DEPOSIT FROM TUBES.

	Scale on Tubes.	Deposit above Scale.
Calcic Sulphate,	50·92	11·60
Calcic Carbonate,	4·18	0·82
Magnesic Hydrate,	14·12	22·21
Iron, Alumina, Silica, etc.,	7·47	9·14
Organic Matter and Oil,	21·06	50·20
Moisture,	1·17	4·23
Alkalies,	1·08	1·80
	100·00	100·00

IV. Deposit from Bottom of Boiler.

Calcic Sulphate,	22·52
Calcic Carbonate,	nil
Magnesic Hydrate,	7·09
Silica, Alumina, and Iron,	31·85
Organic Matter and Oil,	27·95
Moisture,	5·79
Alkalies,	1·80
	100·00

On careful examination of the organic matter and oil present in these deposits, it was found that quite one-half of it was "valvoline," in an unchanged condition, which had collected round small particles of calcic sulphate.

A consideration of these analyses, at first sight, yields no clue as to the cause of the collapse, the scale upon the furnace tops being not only free from oil, but perfectly harmless both in quantity and quality; but, on going more deeply into the question, it is evident that this scale cannot be in the condition in which it was originally formed, as the deposits from both top and bottom of tubes, from the bottom of the furnaces, and from the shell of the boiler, are all rich in oily matter; and it is impossible that, during this deposition, the furnace tops could have escaped whilst all other parts of the boiler became coated with it. Experiments, however, reveal the actions which had been at work and led to the formation of the deposit, and its absence upon the injured portions of the plates.

The pressure at which the boilers were worked was 80 lbs., corresponding to a temperature of 155° C., or

311° F., which is so far below the boiling point of the valvoline that it was evident that it had not distilled over in the ordinary way, and experiments were made to see if it could be distilled in steam at a lower temperature. A retort containing valvoline was carefully heated over a sand bath, its temperature being ascertained by a thermometer, and steam was then blown through it, with the result that at 248° F. or 120° C. the steam became "greasy" and the oil commenced to pass over with it.

This experiment is important, as it shows that in testing the capabilities of a lubricant, the fact that it has a boiling point well above the temperature of the steam is no guarantee that none of it will find its way into the boiler. Having thus entered the boiler, the minute globules of oil, if in great quantity, coalesce to form an oily scum on the surface of the water, or, if present in smaller quantities, remain as separate drops; but show no tendency to sink, as, their specific gravity being ·889 they are lighter than the water, and the difference in gravity is probably even greater at the temperature existing in the boiler.

Slowly, however, they come in contact with small particles of calcic sulphate and other solids separating from the water and sticking to them, they gradually coat the particles with a covering of oil, which in time enables the particles to cling together or to the surfaces which they come in contact with. These solid particles of calcic carbonate, calcic sulphate, etc., are heavier than the water, and, as the oil becomes more and more loaded with them, a point is reached at which they have the same specific gravity as the water, and then the particles rise and fall with the convection currents which are going on in the water, and stick to any surface with which they come in contact, in this way depositing themselves, not as in common boiler incrustations, where they are chiefly on the upper surfaces, but quite as much on the under sides of the tubes as on the top, their position being regulated by whether they come in contact with the surface whilst descending or ascending.

The deposit so formed is a wonderful non-conductor of heat, and its oily surface tends to prevent intimate contact between itself and the water. On the crown of the furnaces this soon leads to overheating of the plates, and the deposit begins to decompose by the heat, the lower layer in contact with the hot plates giving off various gases which blow the greasy layer, ordinarily only $\frac{1}{64}$th of an inch in thickness, up to a spongy, leathery mass often $\frac{1}{8}$th of an inch thick, which, because of its porosity, is an even better non-conductor of heat than before, consequently the plate becomes heated to redness, and, being unable to withstand the pressure of steam, it collapses. During the last stages of this overheating, however, the temperature has risen to such an extent that the organic matter, oil, etc., present in the deposit is burnt away or distilled off, leaving behind, as an apparently harmless deposit, the solid particles round which it had originally formed.

Such a deposit is much more likely to be produced with boilers containing fresh or distilled water, as the low density of the liquid enables the oily matter to settle more quickly, whilst with a strongly saline solution it is very doubtful if this sinking point would ever be reached;

it is evident also that, when oil has found its way into the boiler and is causing a greasy scum on the surface, the most fatal thing that can be done is to blow off the boilers without first using the scum cocks, because as the water sinks the scum clings to the tops of the furnaces and other surfaces with which it may come in contact, so that on again filling up with fresh water it is still found there, ready to cause a rapid collapse.

A very remarkable confirmation of this was found in the case of a large vessel in the Eastern trade, in the boilers of which an oil scum had formed. The ship having to stop some days at Gibraltar, the engineer took the opportunity of blowing out his boilers and refilling with fresh water, with the result that, before he had been ten hours under steam, the whole of the furnaces had come down.

Under some conditions the oil-coated particles coalesce and form a sort of floating pancake, which, sinking, forms a patch on the crown of the furnaces at one particular spot, and under these conditions the general result is the formation of a "pocket."

One curious fact which is worthy of attention is that in most of these oily deposits copper is to be found in considerable quantity. In an analysis of deposit from the furnace of a vessel in which a "pocket" had formed from the above-mentioned cause, the scale showed, as in the case already cited, no reasonable cause for the injury at the damaged part of the boiler, whilst the deposit from the under side of the furnace tubes showed clearly the presence of large quantities of oil matters, which were partly combined with copper :—

Constituents.	Scale.	Deposit.
Calcic Sulphate,	90·354	1·02
Calcic Carbonate,	1·200	nil
Ferric Oxide, } Oxide of Alumina, }	3·200	{ 56·90 { 2·30
Oxide of Copper,	nil	1·90
Magnesic Hydrate,	2·821	1·80
Organic Matter, } Oil, }	1·600	{ 10·46 { 17·84
Sand, etc.,	0·825	7·78
	100·000	100·00

It being a fact that even mineral oils have a considerably solvent action upon copper and its alloys, it is evident that the copper in the oily deposits had been obtained from the fittings of cylinder and condenser. Fortunately this copper is so well protected by oil that in most cases it is extremely unlikely to come in contact with, and deposit on, the metal of the boiler; but if it did, very serious galvanic mischief would be the result.

The next point to determine was the effect which these oily deposits might have in allowing excessive heating of the plates to take place, or in retarding the heating of the water. A clean iron vessel was taken, and a known volume of water placed in it, and heated by a carefully regulated Bunsen flame, the water being raised to the boiling point in ten minntes; this experiment was repeated a second time with the same result, and the vessel was then lined with a coating of deposit found in the bottom of the boilers which had collapsed, and rendered binding by admixture with a small trace more valvoline. This coating was laid on $\frac{1}{16}$th of an inch in thickness, and the former experiment repeated, the same flame being used and the same volume of water taken, with the result that it took fifteen minutes before the boiling point was reached, showing that, even if no damage resulted to the plates from overheating, such a deposit would cause a large increase in the fuel used.

In order to ascertain to what extent extra heating of the plate took place from this cause, Professor Lewes employed a series of substances of known igniting and melting point, raised the water in the various vessels to the boiling point, and then brought the clean bottom of the vessel in quick contact with the test substance, and took the results as indicating the temperature of the exterior of the plates.

Clean vessel . Sulphur did not melt, . . below $115°$ C.$=237°$ F.
Coated vessel . Sulphur melted, . . . above $115°$ C.$=239°$ F.
,, ,, but did not inflame, below $250°$ C.$=482°$ F.
Guncotton ignited, . . . above $200°$ C.$=392°$ F.

So that the $\frac{1}{16}$th of deposit caused with a slow heat a rise in temperature of the plate from under $115°$ C. or $239°$ F. to over $200°$ C. or $392°$ F. It is manifest, however, that the fiercer the heat the more marked will this overheating become, and in the next series of experiments the Bunsen flame was replaced by an atmospheric blowpipe, and the temperature attained, tested in the same way as before.

Clean vessel . Sulphur did not melt, . . below $115°$ C.$=239°$ F.
Coated vessel . Guncotton ignites, . . . above $200°$ C.$=392°$ F.
Tin melts, ... ,, $228°$ C.$=444\cdot4°$ F.
Sulphur ignites, . . . ,, $250°$ C.$=482°$ F.
Lead melts, ... ,, $334°$ C.$=633°$ F.
Zinc melts (just), . . . ,, $423°$ C.$=793\cdot4°$ F.

Whilst, on replacing the atmospheric burner by an oxy-coal gas flame, there was found no difficulty in fusing a hole in the bottom of the vessel, which was made of thin wrought-iron plate, showing that a temperature of $1500°$ C. $= 2732°$ F. had been attained, and it is therefore manifest that, with the fierce heat existing in the boiler furnaces, given an oily deposit only $\frac{1}{16}$th of an inch in thickness, the plates will readily be heated to a temperature at which they are totally unable to withstand a pressure of 80 lbs. of steam, and collapse of the furnace crowns must follow.

The great points to be sought in a good lubricating oil are that it shall be a pure mineral oil, and that its boiling point shall be well above any temperature likely to be attained in the cylinder. Oils satisfying these requirements can readily be obtained, but users of lubricants must remember that, in order to obtain them free from any constituents of dangerously low boiling point, expensive processes have to be resorted to, which must of necessity increase the price of the oil, and that it is useless to expect to obtain a really good lubricant at a low figure.

The great advantages of a good mineral oil cannot be too strongly insisted on, and any lubricant containing animal or vegetable oils to give it body should be unhesitatingly discarded.

The mineral oils are not fats, but hydrocarbons,—compounds of carbon and hydrogen,—and the portions used for lubricants are those left after the more volatile constituents have been distilled off, and they differ widely from animal and vegetable oils, which contain so-called fatty acids, which are liberated from them by the action of superheated steam, and these acids attack iron, copper, and copper alloys with the greatest readiness, forming metallic soaps, which are compounds of the fatty acids, with the oxides of the metals, and so cause serious damage to both boilers and fittings.

Moreover, the lubricating power of a mineral oil increases with its specific gravity, so that the heavier it is the better, and, if properly prepared, it retains its lubricating power at all temperatures short of boiling point. The animal and vegetable oils, in contact with air, especially when heated, take up oxygen, and become gummy and resinous, and gradually so stiff that frequent cleaning becomes necessary, a trouble entirely avoided with mineral oils.

In dealing with the prevention of such deposits, it appears that the most feasible plan would be to pass the feed-water through a long tube filled with clean coke in pieces the size of a walnut, which, acting as a scrubber, would arrest any oil there might be in the feed, and prevent it going forward to the boiler.

It may be mentioned that a filter for this purpose was invented by Mr. J. B. Edmiston, M.I.M.E., of Liverpool, in 1883. Since then the apparatus has undergone alterations and improvement till, in its present form, it is a simple and very effective apparatus. The best testimony as to its value is found in the fact that its use has become essential in the largest mail and passenger steamers, and that it has been adopted by such lines as the "White Star," "Union," "Royal Mail," and in several navies. The filter, of which we give a sectional view (Fig. 48), consists of two rectangular chambers with a semi-circular bottom; each chamber forms a separate and complete filter, and each has its inlet and delivery valves. The two inlet valves are united by means of a breeches pipe with the engine side of the main feed-pipe; and the two outlet or delivery valves are similarly united and joined to the boiler side of the main feed-pipe.

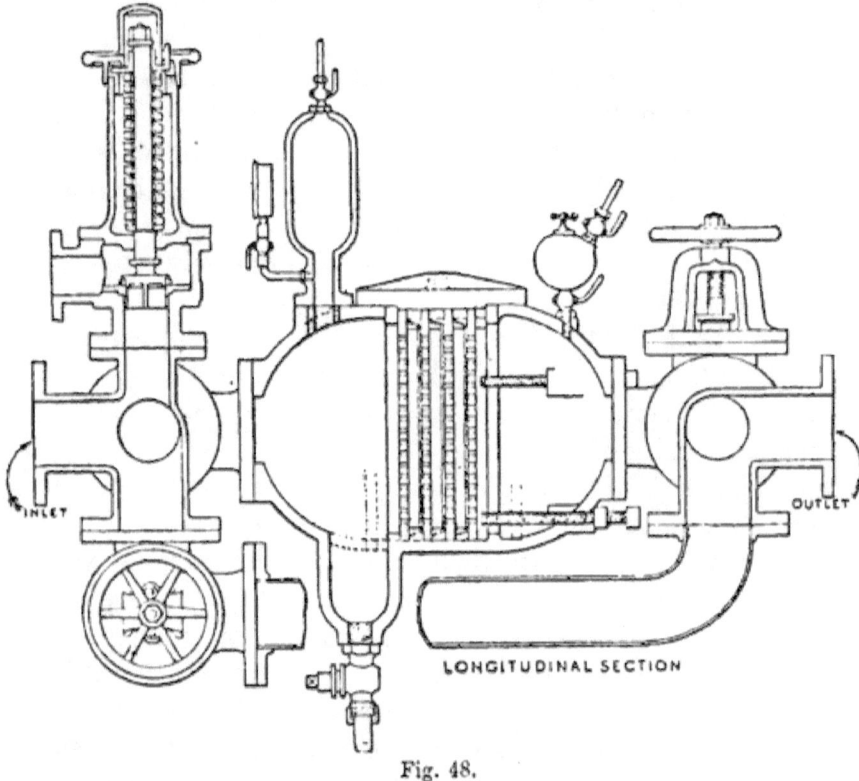

Fig. 48.

There is also a by-pass pipe uniting the two branches of the feed-pipe, thus the feed can pass through either or both filters, or through the by-pass pipe to the boiler. To use the language of electrical engineers, the filters and the by-pass are in "parallel arc," and are placed in circuit with the feed. The filters are formed of perforated iron plates, between which are placed pieces of a specially manufactured cloth or flannel. The perforated plates and flannels are placed in alternate layers, and the whole set up and secured in place by set screws. On the chamber cover are placed the air-chamber, safety valve, pressure gauge, and steam valve. The engineer of the watch is thus enabled at all times to satisfy himself as to the internal condition of the filter. The feed-water is forced by the feed-pumps through the filtering media, with the result that all grease, metallic particles, etc., are mechanically averted, and nothing but pure water is forced into the boiler. When the filtering media become choked, the fact is indicated by a rise in the pressure, as shown by the gauge. In practice, when the gauge shows a pressure of about 2 to 3 lbs. above that required to lift the feed check-valve, it is time to shut off that filter and clean it. Then the other one is used. The length of time a filter will run without cleaning depends upon its capacity, the quantity of feed to be filtered, and the amount of oil used in cylinders. A cargo steamer developing about 1500 IHP., and evaporating about 240 tons water, will require a couple of small size filters, which will run without cleaning for 12 or 14 days; while a full-powered twin-screw mail-boat will require the largest size on each feed-pipe, and the filters will require cleaning about every three days. As, however, the operation of breaking the joints, taking out the saturated flannels, inserting fresh ones, setting-up and making the joints, occupies from 20 to 30 minutes, there is not much time lost. The superior condition of boilers to which the filter is an adjunct is very apparent. There is no scale and no patches of grease, and no deposit of dirty scum, but the interior surfaces are coated with a whitish powder, easily removed by the finger, exposing the bright metal beneath. Recently the steamer *Inchdune*, of the "Inch" Line, arrived in Liverpool, and her boilers were inspected by most of the principal local engineers and boilermakers. The condition was as described above, and excited very general admiration and approval.

It has been already pointed out that the collapse of furnaces from the presence of oily deposits almost invariably takes place in boilers fed with fresh water, and that oil, when it goes into the boiler, floats, because it is specifically lighter than water, and will continue to float until solid particles of calcium compounds coming in contact with it and imbedding themselves in it will so increase its weight, that upon reaching the same density as the water, it will commence to circulate with the convection currents in the water, and being drawn down in this way will readily attach itself to tubes and furnace crowns, causing the damage complained of.

In the case of, say, a transatlantic liner using fresh water only, it would take at least four days' hard steaming to bring the density up to $\frac{1}{32}$, whilst if sea-water had been used the density, being $\frac{1}{32}$ at starting, would never, after the first few days, be much below $\frac{2}{32}$, consequently the oil would remain as a scum, and, unless blowing off was too frequently resorted to, it would never reach the furnace crowns.

Professor Doremus of New York proposes to use sodic fluoride as a precipitant for the scale-forming constituents of sea-water, because a smaller weight of it would be required, and its action would be more rapid and more perfect; and he asserts that the precipitation of the calcium and magnesium salts takes place with very great rapidity, that the sediment shows no tendency to cake or adhere to the sides of a hot vessel, and that it is less bulky than the precipitate formed by soda ash.

The great objection to this hitherto has been its cost, but it is said that this has now been so far reduced as to bring it within the scope of extended commercial use; and, should the demand increase, the price could be readily brought much lower.

There is no doubt that, if the price of sodic fluoride can be so reduced as to enable it to be used for this purpose, it would be of great service, as the fluoride undoubtedly completes its action on the salts of magnesium and calcium more rapidly and thoroughly than the soda ash, and it would be easier therefore to so regulate the quantity added as to remove, say, five-sixths of the injurious constituents from the sea-water without allowing any precipitant to enter the boiler.

In preparing sea-water in the way proposed by Professor Lewes, every precaution must be taken to add slightly less of the precipitant than is necessary to entirely throw down the calcium and magnesium salts, as it is manifestly impossible in practice to guard against small quantities of sea-water finding their way into the boiler either from leaky condensers or else being fed in by the engineer during some emergency; and, if under these conditions any excess of the precipitant were present in the boiler, a bulky precipitate would be thrown down and cause trouble, although it would not bind into a solid scale.

If at any time, however, through a breakdown in the distilling apparatus, sea-water is mixed with the distilled water, a thin and very hard scale of sulphate of lime is formed. This scale is of great interest from the presence in it of carbonate of copper. It is well known that distilled water has a far greater solvent effect upon metals than water containing certain salts in solution, and it is quite conceivable that the distilled water from the surface condensers may attack the brass and copper tubes and fittings, and deposit the copper on the tubes of the boiler, although in only small quantities. It is interesting to note that the green spots due to the presence of copper are all on the under side of the scale, that is, in contact with the metal of the boiler tubes, showing that in all probability the copper had been deposited, as suggested, from the water in the boiler, and its coming in contact with the iron would immediately set up local galvanic action, and thereby tend to produce pitting.

The great difference existing between sea and fresh water would lead one to expect a wide difference in the respective deposits formed from them, and analysis, as given below, of the incrustations formed in the boilers of

Constituents.	River.	Brackish.	Sea.
Calcic Carbonate,	75·85	43·65	0·97
Calcic Sulphate,	3·68	34·78	85·53
Magnesic Hydrate,	2·56	4·34	3·39
Sodic Chloride,	0·45	0·56	2·79
Silica,	7·66	7·52	1·10
Oxides of Iron and Alumina,	2·96	3·44	0·32
Organic Matter,	3·64	1·55	trace
Moisture,	3·20	4·16	5·90
	100·00	100·00	100·00

steamers using fresh river water, brackish water at the mouth of a river, and sea-water respectively, show that with fresh water the incrustation may be looked upon as consisting of carbonate of lime, with small quantities of other compounds: that with a mixture of fresh and salt water the deposit consists of nearly equal parts of carbonate and sulphate of lime; whilst the sea-water gives practically pure sulphate of lime.

The importance of knowing these differences in the deposit formed is very great, because an examination and analysis of the scale which a boiler may contain will enable one to arrive at a sound conclusion as to the treatment it has received during the voyage. Taking, for instance, the case of a ship which uses fresh water both for filling and make-up, it is manifest that on her return to port the scale should be very slight and should consist mainly of calcic carbonate; whilst, if the scale exceeds $\frac{1}{16}$th of an inch and shows a preponderance of calcic sulphate, it is manifest that such scale could only have been formed by sea-water, either leaking in from faulty condensers or being deliberately fed into the boilers.

The presence of sulphate of lime exercises a very marked influence upon the condition and physical properties of the incrustation, as, under the condition in which it is formed in a boiler, it separates in a crystalline form, and binds the deposit into a hard mass, an action which materially assists the presence of hydrate of magnesia.

A deposit consisting of carbonate of lime only, or of carbonate of lime and traces of the oxides of iron and alumina, salt, etc., is separated as a soft powder, which remains suspended in the water for some time, and can easily be removed from the boiler on cleaning it; whilst if sulphate of lime is present the scale is extremely hard, and generally requires the use of a hammer and chisel to detach it from the plates and tubes, an operation which is extremely injurious to them, and tends to shorten the life of the boiler.

The three principal constituents of boiler incrustations may therefore be looked upon as carbonate of lime, sulphate of lime, and hydrate of magnesia, and the causes which lead to their deposition may now be considered.

Carbonate of lime is practically insoluble in pure water, that is to say, it requires more than 10,000 parts of pure water to dissolve one part of the carbonate; but carbonic acid, which is present in all natural waters, when in contact with carbonate of lime converts it into bicarbonate which is soluble, and this forms the so-called temporary hardness in fresh waters. On heating such a water the bicarbonate is decomposed, carbonic acid gas escapes, and the carbonate of lime being insoluble is deposited. Carbonate of lime is slightly more soluble in sea-water than in pure water, but the difference is so small that the above may be looked upon as applying equally to fresh and salt water.

Sulphate of lime is to the marine engineer the most important constituent of boiler incrustation, and its separation as a deposit from water is dependent upon a totally different class of phenomena to those which bring about the precipitation of carbonate of lime. Sulphate of lime is dissolved in water by the solvent power of the water itself, and not through the agency of the carbonic acid, and therefore merely boiling water at ordinary pressure does not suffice to cause its deposition, and for this reason it forms, together with soluble magnesium salts, the "permanent hardness" of fresh water.

Sulphate of lime is much more soluble in sea-water than in fresh water, but its solubility rapidly decreases (1) on concentration of the sea-water, and (2) on the increase of temperature and pressure.

If sea-water be boiled merely under atmospheric conditions, it would be quite possible, by taking care that its density did not rise above a certain point (1·09), to prevent the deposition of the sulphate of lime; but any such regulation of the density is rendered abortive by the fact that pressure and the consequent raising of the boiling point acts upon the sulphate of lime in exactly the same way as concentration and increased temperature, so that in the older forms of boiler, while working even at a comparatively low pressure, most of the sulphate was deposited, whilst in the high-pressure boilers now in use, if the water contains the smallest trace of sulphate of lime, it will be deposited; and as any deposit formed amongst the tubes in such a boiler cannot be easily got at, it is practically impossible to use sea-water in them, consequently they are supplied with water from the condenser augmented by distilled water made in special distilling plant. The question has been raised as to whether if sea-water were mixed with distilled water so as to greatly reduce its density, it would deposit the sulphate of lime; but experiment shows that, even if sea-water is mixed with many hundred times its bulk of distilled water, the minute trace of sulphate present is deposited under pressure when the temperature approaches 284° Fahr. (53 lbs. pressure). So that when, through a breakdown in the distilling apparatus, sea-water has to be used with the condenser water even in very small quantities, a slight scale is sure to be formed.

The loss of heat and waste of fuel entailed by the presence of boiler scale is enormous, and has been estimated by various observers at different values. The latest estimates, however, go to show that $\frac{1}{6}$th of an inch of scale necessitates the use of 16 per cent. more fuel, $\frac{1}{4}$ of an inch 50 per cent., and $\frac{1}{2}$ an inch 150 per cent. additional coal to do the same work. Waste of fuel, however, is not the only drawback brought about by incrustation; the deposit being a bad conductor of heat, the tubes and plates of the boiler soon become overheated and burnt on the outside, whilst rapid corrosion is set up on the inner surface; iron at temperatures far below visible red heat, decomposing water and combining with the oxygen in it whilst the hydrogen is liberated.

Even when sulphate of lime is not present, or is only present to a small extent, as in the deposits from fresh water, the tubes should be allowed to cool down before the boiler is blown off, as, if this is not done, the loose soft deposit of carbonate of lime, when the water has all run off, bakes to a hard tough scale, which acts as a groundwork for further deposits.

The importance of preventing boiler incrustation, and thereby avoiding the enormous waste of fuel and injury which it entails, has not been without its effect upon the minds of inventors, and almost every conceivable substance, from potato parings to complex chemical reagents, have from time to time been patented for the purpose, but they have all more or less failed for marine boilers, because they have often had an injurious effect upon the metal of the plates, or else have produced an enormous bulk of loose deposit, which, although easily cleaned out if the various parts of the boiler were accessible, and if it were only being used intermittently, yet in a marine boiler continuously working rapidly chokes the spaces between the tubes, and brings about an even worse state of affairs than that originally existing.

For these reasons, no treatment of sea-water in the boilers themselves is practically possible, and with high-pressure tubular marine boilers the water must either be condenser water made up to the required bulk with distilled water, as is at present done, or else the condenser water must be augmented by sea-water specially prepared for the purpose, in a separate apparatus, before being supplied to the boilers. In this apparatus the sea-water while under pressure is treated with carbonate of soda, and after rapid filtration is then introduced into the feed-water tank or hot well.

If the engines of a vessel are in perfect condition, they will approximately require 1 ton of water per 1000 horse-power every twenty-four hours, in order to make up the volume of the condenser water to the amount required for the boilers, whilst under most conditions it would be far higher, rising sometimes to as much as ten times that quantity. In fresh-water boilers, that is, boilers filled up with fresh water at starting, and the feed made up from fresh-water tanks, incrustation can to a certain extent be prevented, or, at any rate, diminished, by the addition of substances which will prevent the sulphate of lime binding the carbonate of lime into a hard mass, and these so-called "anti-incrustators" may be divided into two large classes—(1) Those which have some definite chemical action, and (2) those which are purely mechanical in their action.

For preventing incrustations in salt-water boilers, carbonate of soda in one form or another generally plays the principal part. Its action is to convert the sulphate of lime and chloride of magnesia into carbonates, and at the same time to keep the water in an alkaline condition, so as to prevent damage from acids. The hardening effects of the sulphate of lime is thus done away with, and the carbonate of lime precipitated in a soft condition, when it can be readily blown off.

Chloride of ammonia is also used, and when boiled in presence of carbonate of lime will decompose it, forming soluble chloride of lime, whilst the carbonate of ammonia volatilises.

Other alkalis and alkaline salts are also used, and even in some cases, acid mixtures; but these latter, if they do dissolve the scale, are unquestionably fatal to the boiler plates.

There are an enormous number of other organic compounds proposed or in use, their action, if any, being mostly mechanical.

Zinc, in good metallic contact with the boiler plates, in cases where sea-water is used, acts beneficially in preventing corrosion of the plates, but is of no use with fresh water; whilst most other electrical and mechanical applications have met with little success.

Feed-water heaters act beneficially to a certain extent in softening the water supplied to the boiler, and thus reducing incrustation; sometimes nearly 50 per cent. of the carbonate of lime contained by the water having been deposited in the form of mud in the heater.

Corrosion in boilers is of such frequent occurrence, and the causes of it are so little understood, although often discussed by engineers, that it may prove of some service to them if we endeavour to explain the conditions that are considered most likely to result in injury to the plates of a boiler.

When iron is exposed to pure dry air no rusting of the surface takes place, but in moist air containing carbonic acid corrosion rapidly commences and continues at an increased rate until, if sufficient time be allowed for the action, the whole of the metal will be converted into mixtures of oxide of iron and hydrated oxide of iron.

Dry and pure oxygen alone has no action on iron and steel, moist pure oxygen acts only very slowly, and pure carbonic acid when dry has by itself no action; but a mixture of the two in the presence of moisture rapidly sets up rusting of the metal.

When iron or steel is exposed to air it is under these latter conditions, and contact being promoted by moisture the carbonic acid and oxygen simultaneously attack the metal, forming a thin and almost imperceptible layer of carbonate of iron.

The carbonate of iron so formed is at once partially oxidised by more oxygen into basic carbonates, and is finally converted into oxide of iron, whilst if moisture be present in any quantity hydrated oxide of iron is formed. During these reactions the carbonic acid is liberated on the surface of the moist metal, and reacting with more oxygen from the air carries on the process of corrosion, which is now further accelerated by the fact that the hydrated oxide on the surface of the metal is electro-negative to the metal itself, and excited by the presence of moisture and carbonic acid creates a galvanic current at the expense of the metal, and the mass of rust formed being porous, the action continues until all the metal is destroyed and has returned to its natural condition of hydrated oxide.

In fresh water the corrosion which takes place on the surface of iron is due to the same cause, *i.e.* the simultaneous action upon the metal of oxygen and carbonic acid, and if an alkali be added to the water to absorb and neutralise all carbonic acid, corrosion is stopped, and a piece of bright iron or steel may be kept perfectly uncorroded in such a solution.

Sea-water acts upon iron and steel in exactly the same way as fresh water, but with more rapidity, owing to its saline constituents, which also induce a more active form of corrosion by helping to excite galvanic action between the iron in the plates and any other metal present, this being materially aided by want of homogeneity in the plates, by particles of rust, by mill scale, or even by the different amount of work, such as hammering or bending, undergone by different parts of the same plate; and in all of these cases the galvanic action set up causes rapid oxidation of the iron at the expense of the oxygen of the water, hydrogen being evolved.

The main phenomenon of rusting, whether in moist air, fresh water, or salt water, may therefore be looked upon as the simultaneous action of water, carbonic acid, and oxygen upon iron. But besides this action, there are other and local causes in many cases which tend to increase corrosion.

These local causes of corrosion, however, have only to be taken into consideration where they are known to occur, and do not affect the general question.

One of the most important questions of the present day is, does steel corrode more quickly than iron? and in answer to this it may be said that a very valuable research by Andrews has shown conclusively that there is a greater tendency to corrosion on the part of all steels than of wrought iron, when exposed to the action of sea-water. In the course of his investigation, the composition of some of the best known brands of plates was first carefully determined; plates of known weight were then exposed to the action of sea-water in presence of air, care being taken that no galvanic disturbances other than those set up on the surface of the plates themselves should interfere; and the plates were carefully cleaned, dried, and again weighed after various periods of exposure; the loss in weight being taken as representing the loss by simple corrosion.

These tests brought out very clearly the greater tendency to corrosion on the part of steel, and also that the rate of corrosion increases very rapidly with the length of exposure, this latter being due to the fact that the oxides first formed are electro-negative to the remaining metal, so that with the formation of rust, galvanic disturbances are increased on the surface of the plate and tend to increase the rapidity of its destruction.

The probable reason for this increase in the rate of corrosion in steel may be that it is not so homogeneous as wrought iron, that manganese is present in larger quantities, and that the carbon present is partly eliminated during corrosion in the form of carburetted hydrogen which tends to disintegrate the surface.

In marine boilers at present in use, and working at pressures of from 50 to 250 lbs., serious pitting, and, in some places, puncturing of the plates and tubes frequently take place. Before the introduction of hydrocarbon lubricants, such deterioration would have been attributed to the action of the fatty acids liberated by superheated steam from the cylinder lubricants; but as corrosion and pitting have both continued since animal and vegetable lubricants were discarded, and as the trouble appears to increase with the increase in the pressures employed, it is evident that some other solution of the difficulty must be sought.

In short, pitting and corrosion must be due not to one cause alone, but to several acting together, such as the following :—

1. Ordinary corrosion due to the carbonic acid and oxygen dissolved in all natural water, which acts but little when the boiler is in use, the gases being driven off by the steam, but which acts rapidly when a boiler is out of use, more especially when the water is partially blown off and the boiler left wet and exposed to the air.

The simplest cure for this is when a boiler is at rest to fill it quite full with boiled water to which a little lime or other caustic alkali has been added.

2. One cause of pitting in a boiler is the formation of spots of rust from the previous cause, which being electro-negative to the metal rapidly cause it to corrode, and it is this action, and also the galvanic action of traces of metallic impurities, which is, to a certain extent, prevented by the zinc protectors, zinc being electro-positive to both iron and oxide of iron; but for the protectors to be of any use, they must be in metallic contact with the metal of the boiler.

3. The solvent action of distilled water upon metals, which, in modern marine boilers fed with condensed water and specially distilled water, rapidly attacks the plates in places where the presence of a trace of slag or other impurity renders the metal less dense, or gives it a tendency to lamination. This can be prevented by using a saline solution, such as the prepared sea-water already mentioned, for the prevention of incrustation, and by avoiding too frequent blowing off, the action of water on iron rapidly decreasing with an increase in its density.

4. The fact that at a high temperature water is directly decomposed by iron with formation of black magnetic oxide and liberation of hydrogen. This action, if carried out at a dull red heat, is now recognised as the cheapest method of obtaining hydrogen gas. The same action takes place, but more slowly of course, at temperatures below dull red heat, which are often approached on the surface of tubes coated with incrustation, where the moisture present rapidly acts on the metal and ultimately pierces it.

This latter cause of pitting can only be avoided by using either distilled water or prepared sea-water, so as to keep the boiler free from all incrustation.

In a paper upon this subject, which was read by Mr. J. B. Dodds before the North-East Coast Institution of Engineers and Shipbuilders, it is stated that on examination of corroding boilers, those parts serious affected are generally found devoid of the usual hard, protective "sulphate of lime" scale, and in exceptional cases the whole of the boiler is perfectly free from this scale, and the various parts covered with a red or even a black coating of a soft matter, frequently slimy in character.

Much of this matter is found on the upper portion of the boiler in the form of a froth, while the rest is deposited on the tubes, combustion chambers, or settles to the bottom of the boiler like mud.

Attention is particularly drawn to the chemical composition of this deposit: the following analyses may be taken as fair averages of the many samples the writer has examined and analysed :—

	From Bottom of Boiler. Per Cent.	From Top of Boiler. Per Cent.
Ferric Oxide,	65·00	72·9
Calcic Sulphate,	9·02	1·58
Calcic Oxide,	·75	1·38
Magnesia Hydrate,	10·12	8·14
Zinc Oxide,	·75	1·35
Sand, etc.,	1·70	1·2
Oily Organic Acid, combined with the Ferric, Calcic, and Magnesic Oxides,	10·66	10·75
Free uncombined Oil,	2·00	1·25
Water,	...	1·45
	100·00	100·00

In some cases this oily combined acid has amounted to 20 and even 25 per cent.

On examining these deposits one is struck with the curious fact that they contain a very large percentage of magnesic hydrate in an insoluble form, and also with the fact that they contain a very considerable amount of oily organic acid, and that this organic acid is in combination with the ferric and calcic oxides and magnesic hydrates. The presence of this insoluble magnesic hydrate compound at first sight appears unaccountable, and causes us to speculate as to where it comes from, because sea-water only contains magnesia as either sulphate or chloride, both of which salts, especially the chloride, are exceedingly soluble, and are not, as sulphate or chloride, capable of forming an insoluble deposit. Indeed, in one gallon of sea-water only the 98·7 grains of calcic sulphate and the 2·8 grains of calcic carbonate are capable of forming permanent insoluble deposits on boiling and evaporating. All the other constituents are very soluble, and simple boiling and evaporation, to the extent carried on in a steam boiler, would only make the solution stronger without causing their deposition in an insoluble form— that is, supposing no other influences acted on them at the same time. It may be asked, might not the river waters with which the boilers are filled when in port furnish this magnesia? But river waters do not contain magnesia salts as a rule, but are surface waters containing principally calcic sulphate; therefore, this magnesia hydrate of the deposits must be derived from the chloride and sulphate of the sea-water.

There is also the other fact noticeable in the analyses of these deposits, and that is the very considerable percentage of "oily organic acid" present in combination with the ferric and calcic oxides and magnesic hydrates, and this organic acid is derived from the mineral oil used as "cylinder oil" in the cylinders. It is generally stated that these oils are hydrocarbons, therefore not capable of saponification, and that they do not affect metallic surfaces. This is correct as applied to those oils in their natural state, for alkalies do not affect them or form soaps, nor is copper or other metal tarnished or corroded, even after being immersed in them for a considerable or indefinite length of time; but it is incorrect as applied to these oils, when exposed to the influences and conditions existing in the high-pressure cylinder of an engine working at such a pressure that the temperature is higher than the "vaporising point" of the "cylinder oil" which may be in use. All these oils are capable of oxidation, otherwise they would be incombustible; and placed under sufficiently favourable conditions for oxidation, such as very extended surfaces exposed to the action of steam of sufficiently high pressure, and therefore temperature, to reduce a portion of the oil to a vaporous state—and these are the conditions existing in engine cylinders, particularly in the high-pressure cylinder of a triple engine—under these conditions these oils will become in part decomposed and broken up, producing compounds different from the original oil put into the cylinder. These compounds pass forward with the steam, and gradually work their way through the condenser into the boiler, and these compounds, so introduced into the boiler, are capable of combining with bases such as ferric oxide, calcic oxide, or magnesic hydrate, as is proved by the constituents of the deposits already mentioned.

Oil merchants and manufacturers will state that their particular oils have a "vaporising point" of over 600° Fahr.; it has even been seriously contended that the "vaporising point," or that point when vapours become apparent, is at a higher temperature than is the flash point. The writer has examined very many of the standard cylinder oils, and can say that the majority of them, as supplied to ships, vaporise or show vapour when heated to under rather than over 280° Fahr., and flash at under rather than over 450° Fahr.

Seeing that 160 lbs. pressure is not now considered an extraordinary pressure, and that the temperature of steam at this pressure is 363° Fahr., it is not difficult to imagine that much vapour is given off from such oils when used at such temperature, and as this giving off of vapour indicates the decomposition or change of the oil, the amount of such decomposition may be estimated therefrom.

Having pointed out these two peculiarities, let us take into consideration what it is that takes place in a steam boiler, supposing the surfaces of the metal to be unprotected by scale or by artificial means. The salts of the sea-water, especially the magnesic chloride, cause the water to act chemically on the exposed metallic surfaces. This chemical action takes place at all temperatures, and in water of all specific gravities; but the higher the temperature, and the higher the specific gravity or the more degrees the water indicates on the salinometer, the greater is this action. The result of this chemical action is the oxidation of the exposed surfaces of the iron, but more especially of the steel. This oxidation or chemical action at the time produces electricity, as chemical action always does. When this oxidation takes place in a cold solution, the electric tension exhibited is slight, even though the chemical action be considerable. Though this tension does appear slight so far as instruments show it, yet in fact the amount of electricity produced is proportionate to the amount of chemical action; but as both the metal and the water are conductors, and remaining in contact the greater part of the opposite electricities produced recombine and neutralise each other as fast as they are separated. But if this chemical action or oxidation takes place at a high temperature, as in a steam boiler, this recombination does not take place to the same extent, and the salts of the sea-water become electrolysed or decomposed by the electricity, their bases combining with the "oily organic acid" produce the deposits found in the corroding boiler. As these bases of the sea-water combine, and are neutralised by these "oily organic acids," there is liberated an equivalent amount of the acid of the sea-water salt, which helps to still further increase the corrosion.

These reactions take a considerable amount of time to state, but in fact they all take place nearly instantaneously, and this, the writer considers, is the reason why corrosion is attributed to electric action, whereas the electric action in a steam boiler is really due to the chemical action of the water on the metal of the boiler.

In a voltaic couple which may consist of, say, a plate of zinc and a plate of copper immersed in a bath of dilute sulphuric acid, the two plates being connected outside the liquid by a wire, chemical action is set up, and electricity produced in exact proportion to that action. The chemical action causes the electricity, not the electricity the chemical action. The same state of things exists in a steam boiler, there is corrosion or oxidation of the exposed iron or steel surfaces by the action of the sea-water, instead of the corrosion of the zinc plate by the action of the dilute sulphuric acid. The sea-water acting in its own degree as the exciting liquid to produce chemical action and so electricity.

There are two ways of stopping this corrosion, one by rendering the water non-exciting, and the other by taking advantage of a law or fact observed in electricity, which is, that when two elements or metals of dissimilar characters are immersed in a liquid capable of chemically acting on one or both of them, and are at the same time connected together by means of a metallic connection, that element or metal which is most acted on by the exciting medium becomes the positive or corroded element, while the other becomes the negative or inactive element, and so escapes all corrosion so long as they are in metallic contact.

When it is wished to stay corrosion by taking advantage of this electrical fact, the usual method is to employ metallic zinc, being careful to bring it into intimate metallic contact with the metal of the boiler. This will, if sufficient zinc be used, have a beneficial effect; still too much is generally expected from the zinc; engineers expect the effect of these zinc plates, say four of them, weighing in all about 56 lbs., and placed in different parts of the boiler, which will weigh about 30 tons, to influence the whole and every part of the boiler, and to continue to influence it for a period of time. Even if these plates were most elaborately connected in strict metallic contact with the metal of the boiler in its different parts, it is too much to expect from such a quantity of zinc, there being too great a disproportion between the weight of 30 tons and 56 lbs., therefore the areas of its influence must be circumscribed, more especially after being in use a few days, when its surface becomes coated and protected against a great proportion of the corrosion it ought to undergo to enable it to keep its place as the most readily acted on metal, and absorb to itself the chemical action or corrosion which would otherwise attack the iron or steel of the boiler. This idea that the areas of influence of the protective plates are circumscribed to some extent accounts for the fact that a boiler shows signs of corrosion sometimes in one place and then in another; in other words, it shows these signs over areas where the protection influence of the zinc has either been destroyed or too much diminished to be effective.

Fifty-six lbs. of zinc represents ·083 per cent. of the weight of a 30-ton boiler. If from four to five times this amount were used in the first instance, and supplemented from time to time, as it was corroded and rendered ineffective, it would be found that corrosion would be stayed, though there would be considerably more than a proportionate quantity of zinc consumed in a given time than when the smaller quantity was employed. The reason for this larger consumption of zinc being that though zinc in proper metallic contact absorbs all corrosion to itself, it does not destroy or prevent the chemical action or the resulting electricity being formed in the boiler, but rather increases it. As has already been stated, the action of the zinc is simply due to its being the most readily acted on metal, consequently it becomes the positive or corroded element instead of the iron, which would corrode were the zinc not present and in metallic contact. This being the case, it may be unnecessary to point out the advisability of the zinc used being as pure as possible, for any foreign metal it may contain will injure its efficiency in protecting the metal of the boiler, as part of its power would be wasted in becoming positive to that metal instead of to the boiler.

An improved method of applying zinc to boilers which is claimed to be highly effective is that provided by Hannay's Patent Electrogen, which was specially designed for preventing corrosion and heavy scale in high-pressure boilers, and is really the outcome of an extensive series of experiments conducted by J. B. Hannay, Esq., F.R.S., *Ed.*, F.I.C., F.C.S.

These experiments demonstrated that the varying conditions of the different parts of the metal in the boiler — sometimes originating in difference of composition, but more generally arising from difference of temperature — cause galvanic currents to be set up, resulting in the corrosion of such portions of the metal as have become *positive*. This is quite in accordance with the well-known fact that, where two metals are placed in a solution which can act upon both and then joined by a wire, that metal which is more electro-positive than the other alone suffers from corrosive action; and so long as the positive metal is metallically connected with the other metal, no action can be exerted on the latter; and it is this natural law, forming the principle of the galvanic battery, which is taken advantage of in applying the electrogen to boilers.

The tubes and internal shell of the boiler are made the *negative* electrode of a galvanic couple, and by thus preventing any part from becoming *positive, corrosion is rendered impossible;* whilst the hydrogen evolved from the negative surface, besides being the best preservative of iron known, causes the scale to fall off before it can do any harm.

It has been shown conclusively in the Report of the Admiralty Commission on Boilers (1880), that while ordinary zinc, when applied by mere clamping or bolting, may yield some protection to the iron, it has the effect of causing the scale to adhere more firmly; but the current set up by the electrogen is of such intensity that hydrogen is set free underneath the scale, which, whenever it becomes thick enough to be impervious to hydrogen, blisters off, leaving the iron with a thin protective scale only, in this way keeping the scale forming and re-forming in thin coatings, and allowing the fuel to do its full work.

For high-pressure boilers the electrogen is especially valuable, and the wire connections being maintained by means of screwed studs of a special pattern, which insure perfect contact, and are easily adjusted for renewals, and all other portions being manufactured with the greatest care, it can now be guaranteed to give thorough protection so long as an ounce of the zinc remains.

To those who are not familiar with this invention, the following directions as to fitting may be useful:—

To fit a double-ended boiler, fix the holders, B, as shown in Fig. 49, by fastening the arms to upper row of tubes, in positions as follows:—Where six electrogens are required: *at front*, one holder in *right* wing, the second in *right* centre space, and the third in *left* centre space. Place electrogens in holders, drill holes, screw in studs, and fix wires as shown in diagram. At back fit in *left* wing and centre spaces.

Where four electrogens only are preferred, the holders should be placed in the centre spaces, wires being led as shown in diagram. The wire in left centre space is spliced, one end being fixed to *wing* furnace side, and the other to *centre* furnace side; the single wire to combustion chamber top. The wire in *right* centre is spliced at top, one end being fixed to each combustion chamber; the single wire to *right* wing furnace side. The fitting is simply reversed at other end of boiler, but

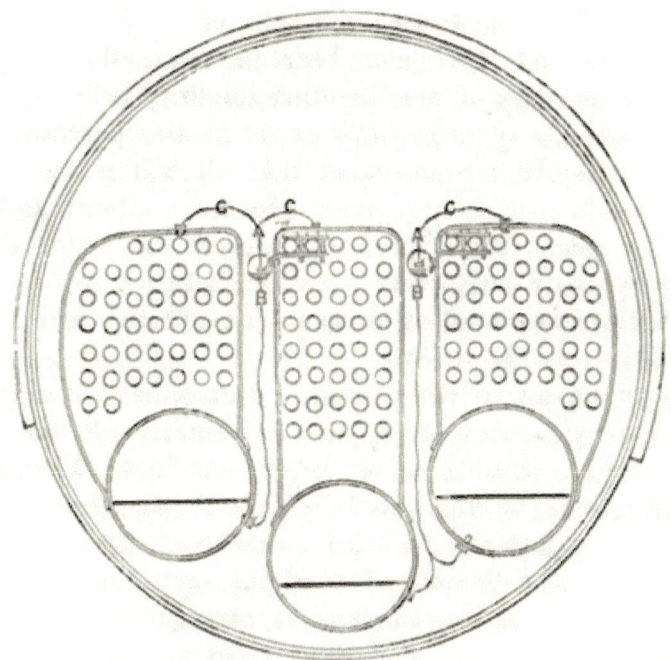

Fig. 49.

great care must be taken to see that the studs are fitted tight, and wires thoroughly connected to the stud.

To solder the wires efficiently, a space on the metal, 3 inches long by 1 inch broad, should be thoroughly cleaned with a chisel, then heated and tinned with a heavy soldering bolt, about 8 lbs. in weight, so that the solder may thoroughly run into the iron surface. Then, with a second soldering bolt, about $2\frac{1}{2}$ inches of the wire, after being cleaned and brightened, must be firmly soldered thereon; and should a wire become broken, the two ends, besides being brightened and twisted, must always be *soldered*, as the mere twisting does not insure electrical contact. The solder used must be of the finest quality (pure lead and tin); and bear in mind that if the smallest quantity of zinc or other impurity is present, it will *render the solder perfectly useless* for this purpose.

It is highly recommended that all boilers be fitted with studs, particularly high-pressure boilers; and it may be added that, should any part of the boiler show itself specially subject to corrosive action, a wire from the electrogens applied to the part affected will most speedily arrest the decay.

With regard to their general management, it may be said that when they are applied to a boiler for the first time it is advisable to scum off and blow through a small portion of the water daily for the first few weeks, in order to clear out the loose particles of rust, dust, and corroded matter thrown off from the surface of the iron by the action of the electrogens, and which, if allowed to accumulate, will give the boiler an unsatisfactory appearance, and more especially should this procedure be adopted with absolutely new boilers, so as to thoroughly clear out the rusty scale which is upon new iron, and which comes off through use; for this iron scale, being a non-conductor, is thrown off more rapidly by the action of the electrogens. At the same time, it will probably be disintegrated and oxidised to a light red pulverent body, which, if not cleared out, is apt to get the boiler into a dirty condition, and prevent the deposition of a thin lime scale. After this red matter has been removed, the boilers will gradually assume the light buff colour characteristic of the use of electrogens.

When a boiler is fitted with them for the first time, the electrogens become exhausted somewhat speedily owing to the greater demand made upon them, but they usually last four to eight months. After this first action, a comparatively feeble power is sufficient to preserve the surface of the iron, and they will remain effective for an average of six months, or longer, after which they should be replaced by fresh ones.

It may also be mentioned that, as the presence of some salt in the water is necessary to bring the electrogen into proper activity, it is recommended that the boilers be filled up with clean sea-water, which may be done with perfect safety, as the boilers are fully protected by the action of the electrogens.

The electrogen is found to act most efficiently at densities ranging from 2 to 8 ounces per gallon, and its first action, when corrosion has been going on in a boiler, is to throw off the oxidised iron, clearing out the decayed matter from the pitted cavities; and while this is going on, the boiler may have a red appearance, but this soon passes away after the iron has become freed from the incrustation of rust.

The electrogens should, of course, be examined from time to time, to see that the wire connections are secure, and to thoroughly clean off the crust of oxidised zinc, otherwise their effective action will be impeded.

It would certainly seem that the most logical method of preventing corrosion is to make the water non-exciting or incapable of acting chemically on the iron or steel of the boiler; thus the cause is at once attacked, whereas the other method only deals with the effect, and there is, moreover, avoidance of the great difficulty in making and maintaining the metallic contacts, owing to corrosion at the point of juncture, or the breaking of the contacts from other causes, and such imperfections can only be remedied when the boilers are opened. In those methods which aim at destroying the corrosive or exciting power of the sea-water, the protective agent is added either at the condenser or hot well from time to time, and in greater or less quantities as desired. There are several ways of making or causing the sea-water to be non-exciting, and there are many compounds offered as meeting all the requirements; but care should be taken not to put into boilers any compound which contains a constituent that of itself is capable of combination with iron, or which contains a constituent that can by any means be made to furnish compounds capable of such combination, because it should be a *sine qua non* that the protective agent should be in itself harmless. Lime preparations added to the water are beneficial, their action being to keep the surface of the boiler always coated and thus protected from corrosive action. This subject has occupied so much of Mr. Dodds' professional attention that it has caused him to take more than ordinary interest in the solution of the problem, apart from any commercial consideration of the question, although he felt it would be a great advantage if some reduction could be made in the present costly application of zinc, and he is strongly of opinion that a basic solution of zinc would effect this economy, but, after all, he assumes that the vitality of the boiler is the first consideration.

The great advantages in the use of such anti-corrosive or anti-exciting compounds is that they can be introduced in small quantities at stated intervals. They render the water non-exciting, and diffuse themselves through all parts of the boiler, thus protecting all parts equally.

In treating of corrosion, mention of that special kind generally known as pitting has been omitted. This pitting is occasioned by the same causes as induce the more general corrosion, but these causes are intensified and accelerated by two other influences, which tend to concentrate the effects of such corrosion by rendering it very local instead of general. These influences are rust or iron scale and variations of temperature. Rust or iron scale is frequently, indeed generally, in the form of "magnetic oxide of iron," and when the metal of the boiler, especially if it be steel, is acted on chemically by the sea-water, and whilst in intimate contact with this oxide, such chemical action induces electricity, the oxide and the metal in its very immediate neighbourhood constitute a voltaic couple, the metallic iron or steel being the most readily acted on becomes the corroded or positive element, while the oxide becomes the inactive or negative one; this couple induces a current of electricity having only a very local influence, thus concentrating the action on that limited portion of the iron or steel which has become the positive or corroded element through the influence of the oxide or scale, instead of allowing that action to expend itself more generally over a larger area.

Variation of temperature affects more particularly the question of the very serious and dangerous pitting observable on the sides of the furnaces. In cases where two portions of even the same plate of iron or steel are subjected to unequal temperatures when immersed in a liquid capable of chemically acting on them, these two portions become virtually two different metals so far as molecular arrangement is concerned, and are capable of forming a voltaic couple; the more highly heated portion being the most readily chemically acted on by the seawater becomes the positive or corroded element, while the less highly heated portion, being also the less liable to the chemical action, is the negative or inactive one. Thus, when through any physical or structural cause one part of the metal becomes more highly heated than another part,—and portions of the furnaces and combustion chambers are very liable to this, especially along the fire line,—this more highly heated portion becomes positive to the less highly heated portion, and thus concentrates on itself all the corroding or chemical action which would have diffused itself more generally over the whole surface had the temperatures been equal.

To counteract or stay this pitting is much more difficult than it is to stay the general corrosion. In the case of general corrosion there is a general cause which may be met by a general cure; but in the case of pitting there are several causes, each perhaps similar, but yet each requiring to be separately neutralised and overcome. Each case of pitting being due to a local and not a general cause, in any endeavour to effect this cure by means of metallic zinc it will be necessary to remove this cause and bring the part affected into intimate metallic contact with the metallic zinc. Bringing the zinc into this contact with these parts will very greatly increase the consumption of zinc, seeing that these parts are so prone to chemical action. Supposing the causes, such as rust or iron scale, and the variations of temperature to be removed, this increased use of zinc will be effective. But though it may be possible to remove the rust, it is not so possible to do away with the variations of temperature; therefore the best method of effecting a cure of this pitting would be to strike directly at the cause by rendering the water non-exciting. By this means the rust and the variations of temperature are rendered innocuous.

In conclusion, the writer would point out that prevention is better than cure; and that if it is desired to keep a boiler in good order, certain precautions must be taken. Firstly, great care must be taken in the selection of cylinder oils, and only those must be used which have a vaporising point at a higher temperature than the temperature of steam at the pressure at which the boiler is worked. The statement given of the vaporising points of these oils must not be taken for granted, even with a certain brand or make, but each particular lot supplied must be equal to the sample, and bear out all the statements made respecting it and which influenced its purchase. Such supervision will always give a good return for the trouble, because whether it is cylinder oils or any other goods that are sold, and which are not subjected to this supervision, it may be taken for granted that the lowest quality accepted without serious complaint or rebate in price will eventually become about the highest quality that will be supplied.

Secondly, the boiler must be worked with the greatest amount of regularity practically possible on board ship, and the specific gravities of the water kept as regular as possible. Samples of these waters may be taken at stated times during the voyage, and of deposits whenever opportunity offers; these samples to be kept for examination when necessary. The taking of these samples answers two good purposes—one is, that if the boiler should happen to show signs of corrosion, these samples will enable the cause of such corrosion to be traced; the other is, that by taking samples systematically and for a certain purpose attention is thereby drawn to the boiler and a certain interest created, which induces regularity in working and general treatment. All measures which have this effect are most valuable, and cause a very material reduction in the wear and tear of the boiler.

CHAPTER IX.

Management of Working Surfaces—Crank Shaft, Crank Pin, Pump Lever Brasses—Liners—Chipping Brasses—Bearing Surface—Hints on Adjusting Brasses—Care of Built Crank Shafts—Guide Bars and Shoes—How to Lubricate them—The Thrust Block—Causes of Trouble and their Remedies — Adjustment of Tunnel Bearings — Warming up Engines—Making Ready to Start—Precautions under Way—Sweeping Tubes—Cleaning Fires—Blowing down Boilers—Treatment of Ashes—Fuel Economy.

It would be difficult to lay down any fixed rule regarding the management of triple and quadruple expansion marine engines, as each set will be found to have certain peculiarities which will require special treatment; but at the same time there are a number of well-defined points which ought to be noticed, and the remarks upon which will apply with equal force to all engines coming under the above description.

There is perhaps nothing more annoying to an engineer when at sea than to be troubled with warm bearings or with engines which, judging from the noise they make, bear a strong resemblance to steam hammers.

No doubt the most of such troubles can be obviated in exact proportion to the skill and energy displayed by the engineers of the ship; but, on the other hand, the most expert and industrious engineer can do extremely little to remedy defects due to badly designed working parts, such as insufficient bearing surfaces.

Crank shaft and other journals which revolve in their bearings will be found as a rule less liable to give trouble than those which merely vibrate, such as connecting-rod top-ends, eccentric rod-ends, air-pump links, centre gudgeons, etc.

These vibrating bushes should never on any account be eased away at the sides, as many instances have occurred of broken top-end brasses, which can be attributed to no other cause than having been chipped away at the sides, for about $1\frac{1}{2}$ inches from the face of each brass,

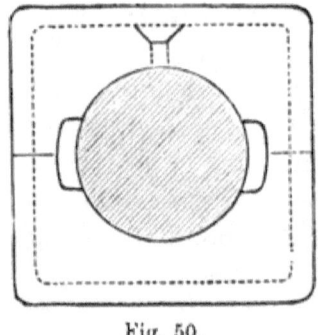

Fig. 50.

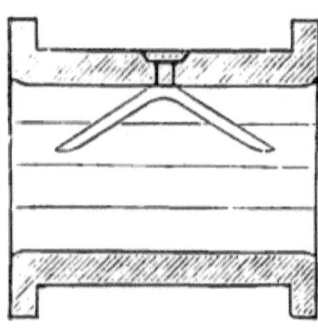

Fig. 51.

and about $\frac{1}{4}$ inch deep, the appearance of which is shown at Figs. 50 and 51, which, with the following illustrations have been taken from a paper on this subject which appeared in the *Engineers' Gazette* for October 1889.

Brasses which are treated or rather maltreated in this manner are naturally bound to give trouble. Under normal conditions the gudgeons have a tendency to wear oval; what else can be expected but that this evil will be intensified when one-third of the bearing surface of the brass has been cut away from the sides.

It is said that it is the practice of some engine-builders

to cut away the gudgeons at the sides, as in Fig. 52. Should this be done along the whole length of the journal side play will be permitted unless carefully followed up by the brasses as wear takes place, in addition to which it will allow the lubricant to escape from the journal instead of remaining on it and providing for the lubrication of the lower brass. In cases where top-end or other similar brasses will not run cool it will generally be found that the cause is due to either oval gudgeons or else imperfect oil ways.

In the latter case the oil channels may, for instance,

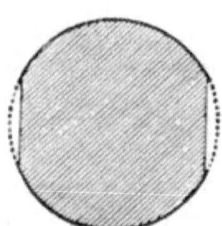

Fig. 52.

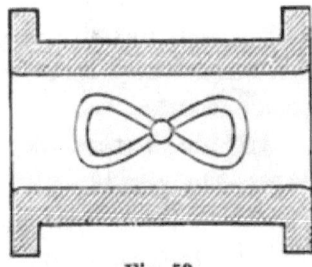

Fig. 53.

have been cut from the oil hole in the crown of the top brass down into the recess at the sides, as shown in Fig. 51. If so, the oil is virtually running down a pipe, the end of which leads outside the bearing.

One method of overcoming this objection is by cutting the oil ways of all vibrating journals in the form of the figure eight, as shown in Fig. 53, which will retain the oil as long as possible on the crown of the brass, and there need then be no fear of it not lubricating the bottom brass, provided there is no escape at the sides as already mentioned. One fertile cause of broken bushes, especially amongst the smaller ones, such as cross-head, pump lever and winch brasses, is the cutting of a deep oil way in line with the axis of the journal, or, in other words, along the crown of the brass, as shown by the dotted lines in Fig. 54. To avoid this, the best plan is to cut a curved oil way in the manner shown by the black lines in the same figure.

Another most objectionable practice, and one very frequently adopted by a certain class of engineers, is to chip away the edges of the bushes as much sometimes as $\frac{3}{8}$ or $\frac{1}{2}$ inch, and then make that amount up by thin liners. This method saves the brasses being taken adrift when adjusting them, but it is a lazy one, and this is probably the reason why it is so extensively resorted to.

All the bushes except, perhaps, the crank-pin bushes, should be run brass and brass, without any liners in. This, of course, necessitates the various brasses being taken apart every time they require to be readjusted; but that is rather an advantage than otherwise, for the different parts can then be thoroughly examined and cleaned.

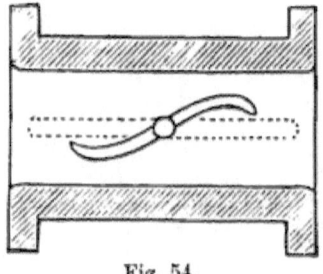

Fig. 54.

It very often happens that the easiest brass to take out, such as the bottom brasses in top and bottom ends, and the top brasses in eccentric rod-ends, air-pump lever gudgeons, links, etc., are chipped away when adjusting them, while the other half is allowed to retain its original thickness. This should not be the case. The brasses should be carefully callipered, and the thickest half reduced by chipping or filing when required for purposes of adjustment.

Regarding the main bearing and crank-pin brasses, they may and ought to be eased a little at the sides of both the top and bottom halves of the bush, to within, say from 1 inch to $1\frac{1}{2}$ inches of each end of the brass; this recess then forms a reservoir for the oil, and hot journals, through want of oil, should then be of rare occurrence.

The usual method of easing them is to separate them with a liner about 1 inch thick, after the brasses are bored out to the size of the shaft; then the boring tool being made slightly larger than before will remove the metal at the sides, as shown in Fig. 55, without touching the top or the bottom. The reason why these brasses are better to be eased at the sides is on account of their tendency to close in, especially with the old style of pockets; with the modern half-round brasses, when there are any strains on the foundation causing the pocket to close in on the brass, they will not be felt so much as with the old rectangular sides.

Fig. 55.

On no account should tin liners be put under the bushes when lining up a shaft, for if the sea-water should get to them the probability is that they will soon become completely corroded away, and the bearing will then be that amount too low.

With a shaft truly lined up, it is very important to have the brasses nicely adjusted, for a bearing will work warm at times if left even a very little slack; but a great deal, of course, depends upon the kind of metal of which the bushes are made and the quality of the lubricant used, keeping always in mind that a shaft may be run finer with thin oil than with thick.

To show the importance of having a bearing properly adjusted, an exaggerated illustration is given in Fig. 56, showing the shaft practically running or rather bearing on a line, but it will readily be seen that as the brass is closed in to fit the journal the bearing surface is increased.

As the total friction of a journal remains constant under ordinary circumstances, any decrease of bearing surface must of necessity increase the friction per square inch upon the remaining bearing surface, and consequently increase its liability to heat. Special care and attention are necessary in the case of built crank shafts. The crank pins especially require to be well watched in order to prevent them heating. It is extremely probable also that after a few expansions and contractions the crank pin will become loose in the web; therefore, when the pins show the least sign of wearing oval, they should be carefully trued up, and this will save future trouble.

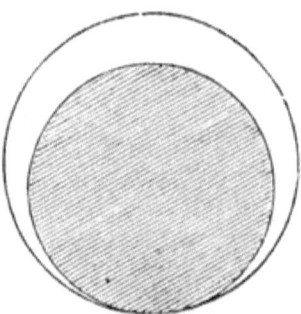

Fig. 56.

It is only on very rare occasions that much trouble is experienced with the guides of modern marine engines, more especially those that have water circulating at the back; but very often a large quantity of oil is wasted by them, when the oil ways happen to be cut in such a manner that the lubricant escapes from the surfaces before having done its duty. Fig. 57 shows one example of this. The guide is fitted with three oil pipes at the top, and it can easily be seen that the oil runs down into the groove A, from whence it escapes at both ends and trickles down the sides of the guide into the bilges, this evil being aggravated, of course, when the ship rolls, in which case the sides of the columns, instead of the guide, will get most of the lubrication.

A good arrangement for the lubrication of guide shoes is shown in Fig. 58. Two or three grooves are cut about $\frac{1}{4}$ of an inch deep, $\frac{3}{5}$ of an inch wide, and in length about $\frac{2}{3}$ the width of the shoe, these being connected by smaller grooves cut diagonally as shown in the figure. The number of grooves, of course, will depend upon the size of the shoe, and for smaller ones two will probably be found sufficient.

By adopting this method a small portion of the bearing surface will undoubtedly be lost, but the improved lubrication that results therefrom will more than make up for it. Where there are grooves cut across the guide, this plan would not answer so well,

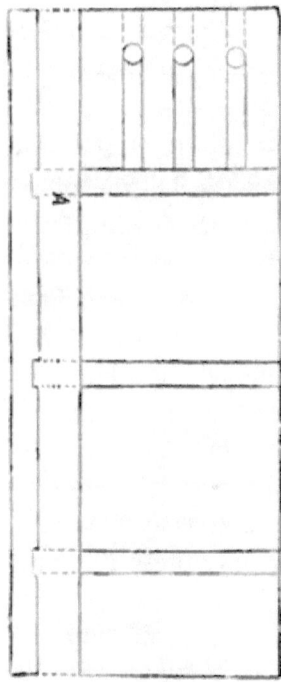

Fig. 57.

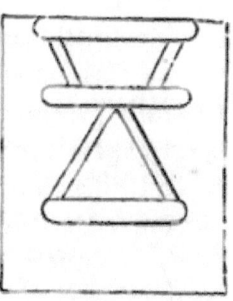

Fig. 58.

the idea being to keep the oil in the grooves cut in the shoe. But even these might be improved by filling up the ends of the grooves with a small piece of metal about $\frac{1}{4}$ of an inch wide, which would at least have the effect of keeping the oil on the guide face, and preventing it from running out of the ends of the grooves.

Drip boxes should always be fitted on the bottom of the guide bars, and a metal comb on the bottom of the guide shoe to dip into the boxes at every revolution of the engines. A good plan is to make a second comb of leather, of about $\frac{3}{4}$ the length of the other, and have it fitted between the metal comb and the guide bar. This simple arrangement often enables guide bars, which had previously given considerable trouble, to be run without water. It may be mentioned that these leather combs cannot be safely used where the grooves are cut square across the guide bars, as they would be liable to get into such grooves, and very likely do more harm than good.

The thrust-block or bearing is also sometimes a cause of trouble. This may occur in various ways—first, through insufficient bearing surface. This is a defect that is only met with in the old-fashioned thrust, and not in the modern horse-shoe design. It cannot, of course, be remedied; but it can be minimised by seeing that all the rings are bearing equally, and that each ring is properly lubricated. Trouble may also be caused by the oil pipes and channels becoming choked. This can, of course, be prevented by frequently washing the rings, pipes, channels, and worsteds with hot water—in fact, thorough cleanliness throughout must be the rule if a satisfactory running engine is desired. A thrust may become heated when it is turned into what is often erroneously termed a thrust bearing—that is to say, if the rings bear on the bottom they are almost certain to give trouble. The obvious remedy for this is, of course, to take sufficient liners out from underneath the thrust-block to prevent the rings from bearing on the bottom. The modern horse-shoe thrusts are a great improvement on the old style, and give very little trouble; at the same time they require to be kept extremely clean, in the best of order, and frequently and carefully adjusted. Care must also be taken not to let the collars bear on the bottom of the block, if liable to do so, otherwise even they will be liable to heat; but it is very seldom they are so constructed as to render this at all likely.

The tunnel shaft bearings should never be let too closely together, especially when they are on the top of ballast tanks, as the filling of the tanks or the loading or discharging of the ship's cargo may strain the shafting and throw it out of line, in which case it is almost certain to heat unless it has sufficient freedom allowed to enable it to accommodate itself to the altered conditions, and to distribute the strain along its entire length.

Before getting under way, allow plenty of time for warming the engines thoroughly. Open all drain cocks, and when all the water is drained out close them again, then open the stop valves as soon as the steam begins to show, at the same time putting the slide valves into such a position that the steam can pass freely into the steam chests and cylinders. The auxiliary valves and steam-jacket cocks should also be opened, so that every part may become gradually and equally expanded. Before moving the engines, carefully examine them all over, in order to see that all the working parts are free to move, and that nothing is lying about which would be likely to fall amongst the machinery when in motion. If possible, give the engine a few turns before starting—this will warm all parts. The donkey should also be started in good time to circulate through the condenser, in order to keep it cool. Assuming the cylinders to be jacketed, keep them clear of water, and only use a small quantity of steam when under way. The scum should always be taken off the top of the water in the boiler before starting, and as there is always a certain amount of grease and dirty water in the cylinders and condenser when in port, the feed-water ought to be allowed to discharge into the bilges for a short time after starting and before permitting it to enter the boilers. Make it a point to have the engines in perfect readiness for starting exactly at the time ordered; by this means undivided attention can be given to their working. Keep the stop-valve full open, admitting the proper quantity of steam by the throttle valve, and if it is not required to work the engines up to their utmost powers, regulate the speed by means of the expansion valve or link motion. To obtain the best results, the engines should be worked at about 80 per cent. of their full power, as the fires do not require to be forced so much, and consequently the fuel has a higher evaporative efficiency. Feed-heaters are now almost invariably fitted to triple expansion engines, but where this is not the case a drain pipe from the intermediate steam chest led into the hot well will be found very useful and economical, if it is opened just sufficiently to take away the accumulation of water that may have gathered through condensation; it will also save the packing in the valve spindle glands, and heat the feed-water without interfering with the vacuum, provided the air-pump is fitted with metallic valves. All expansion joints should be frequently examined, and care taken to see that the steam pipe works freely in the gland, for if the packing becomes hard and solid it ceases to act as an expansion joint, and there is a great danger of the pipe suddenly fracturing when the ship is straining heavily.

The steam steering gear and the donkey engines should be made to exhaust into the main condenser when the ship is under way, so that there may be as little loss of steam and water as possible when they are working. The condenser should be overhauled occasionally, and the outside of the tubes treated to a strong solution of caustic soda; this can be best applied by means of a syringe when in harbour, and the condenser should afterwards be washed out with hot fresh water. Be careful to keep the bilges perfectly clean, and the strums and bilge injection free from dirt; this will save many an hour's hard work when at sea and in bad weather, in the way of clearing choked strums, pipes, pumps, etc.

Before relieving the watch everything about the engine and boilers should be thoroughly examined, and the water-gauge should be thoroughly and carefully tested by the engineer coming on watch in presence of the engineer who is being relieved, and at any other time that the water in the glass appears disturbed or lower than usual. This can best be done by closing the cock on the water space of the boiler, closing also that at the bottom of the gauge column, when, upon opening the drain cock, if steam blows out it shows that the steam passages are clear and that the glass is showing the

P

correct water level; on the other hand, if steam does not blow out, the steam passages are either closed or choked.

It is hardly necessary to warn engineers against the danger of allowing the water to get too low in the boiler, but it must be borne in mind also as of some importance that it should not be allowed to get too high, for this would necessarily diminish the steam space, thereby causing an irregular supply of steam to the engines; and there being a greater quantity of water, it would require more fuel to keep it up to the temperature of the steam in the boiler, besides, in many cases, it would seriously increase the boiler's liability to start priming. A handy rule for finding the safe working water level is to multiply the diameter of the boiler *in feet* by ·3, and this gives the height of the water *in inches*, to be maintained above the highest part of the combustion chamber. This can, of course, only be ascertained when in port by going inside the boiler and measuring the distance between the top of the combustion chamber and bottom of gauge-glass.

For the sake of clearness the following numerical example may be given:—Suppose the diameter of the boiler is 15 feet, then 15 × ·3 = 4·5 inches, or the depth of water to be carried on top of the combustion chamber.

Should anything be found out of order when going upon watch, insist upon the engineer who is going off watch putting it right before leaving. There should be a proper place for all tools and spare gear, and when overhauling in port see that all these are put back into their respective places before leaving work. Should there be any missing, look for them at once and until they are found, as it may save the risk of an accident when getting under way through their being left amongst the machinery. In frosty weather, when not under steam, take the precaution to drain all water from cylinders, jackets, pumps, condenser, and other places where it is likely to lodge; keep the temperature of the engine-room above freezing-point, and this will obviate the danger of pipes bursting, or of the cylinders or other parts of the engines cracking or giving way when getting under steam again.

Take every advantage of fair winds, for with a good stiff breeze and all sails set the engines can be linked up to more or less advantage; depending, of course, on the force of the wind and the class of ship. When steaming against a strong head sea, and the ship plunging heavily, it has been found a good plan to link up sufficiently to reduce the speed two or three revolutions; the ship will travel about as fast, the excessive strain on the engines will be reduced, and coal will be saved.

Considering the high pressure of the steam now used, it is evident that increased care and attention must be bestowed upon the boilers. The life of a boiler depends upon the quality of its materials, its construction, and its treatment; and as cleanliness, both within and without, is the most important factor in the preservation of a boiler, it should be kept as free as possible from dirt and scale, and frequently subjected to careful examination, noting any signs of weakness, corrosion, or defects, and having them rectified, or the parts protected at the earliest possible moment.

Dirty tubes have the effect of diminishing the area of the passage through which the products of combustion necessarily pass, and consequently injure the draught to a corresponding extent. Soot also is a very bad conductor of heat, and considerable loss may thereby ensue from dirty heating surfaces; care should therefore be taken to keep them as clean as possible, by sweeping the tubes regularly when under steam, and thoroughly cleaning the combustion chambers and flues when in port. Even if steam can be easily maintained, it is more economical to keep the tubes clean than to allow them to become choked. A good plan is to have the smoke-boxes fitted with small sliding doors at the bottom, and by occasionally opening these and cleaning out the soot the tubes will run much longer without becoming choked —the draught being usually sufficient to draw the soot clear of the tube ends.

The boiler bottom which is exposed to the action of the moisture and gases rising from the bilges should be kept perfectly clean, and protected by two or three coats of red lead or oxide of iron paint. All leaky seams should be caulked immediately they are noticed, care being taken not to spring the plates at the edges by using too thin a caulking tool.

When the main boilers are not fitted with any special arrangement for circulating the water while raising steam, the lower fires should be lighted at least twelve hours before steam is required, lighting the wing fires about four hours after the lower ones. By this means the water will become gradually heated, the expansion of the plates be more equal, and this will tend greatly to prevent leakage. Forcing the fires in order to get up steam quickly turns a new boiler into an old one sooner perhaps than anything else, as it leads to leaky seams and tube ends, and to injurious straining of the plates.

Among the various causes which make boilers difficult

to steam, the principal are inferior fuel and inefficient stoking. Boilers being designed to evaporate a certain quantity of water with a given quantity of a good quality of fuel, it is evident that if the fuel burnt is not up to the standard, a deficiency of steam must be the result. That shortness of steam is due also to inefficient stoking is so well known that it only requires to be mentioned. Firemen should be instructed to break all large lumps of coal into pieces of a nearly uniform size before firing, as by this means a more equal and effective combustion will be maintained. A good deal of care is also required in cleaning the fires properly. They should never be pulled right out. It is much better to clean one side at a time, as there will be less danger of the cold air injuring the plates, and of quantities of half-burnt coal being pulled out of the furnace.

As economy in the consumption of fuel is of as much importance as economy in anything else, engineers should strive, by every means in their power, to prevent waste of coal. After the fires are cleared and before the ashes are hoisted out of the stokehold they should be carefully sifted and all coal and cinders taken out and only clinkers and small ashes that will not burn thrown overboard. This may be considered by some engineers to entail so much trouble as to be not at all commensurate with any saving that might be effected; but there are cases, however, on record where ashes have been thrown away that were proved by analysis to have contained not less than 45 per cent. of useful fuel. This would in such cases mean a saving of probably 5 or 6 per cent. in the consumption of fuel, a result by no means to be despised.

To overcome the objection, that it is too troublesome

to separate the unburned coal from the clinkers and worthless ashes, various mechanical appliances have been devised.

The most noteworthy of these is probably the Ashes Washing Machine invented by Mr. Leoline A. Edwards, which separates the refuse from the fires into three portions, viz. :—1. Unburnt fuel, called "breeze" or cinders, which can be mixed with coal and burnt over again, or used for blacksmiths' fires, being the finest material procurable for that purpose; 2. Fine dust, useful for builders in place of sand; 3. Clinkers, for making roads, paths, concrete, etc.

In the complete machine, that is with an extra sieving arrangement and automatic feed (which are not shown in our illustration, Fig. 58), the action is as follows :—Over a pit there is fitted a grating of bars 2 inches wide, the stoker tips his ashes over this grating, and all that passes through is carried up by means of an elevator into the machine. The large clinkers which remain on the top of the grating are brushed off, leaving the grating clear for the next barrowload of ashes. On entering the machine, the dust is removed at once by falling into a screen. The clinkers and breeze then enter the machine, to be washed and separated. The machine consists of a tank or compartment kept full of water, the ashes or breeze to be cleansed resting on a grating which is covered with perforated copper plate allowing a free passage for water, at the same time preventing the fuel or breeze falling through. The separation is effected by an agitator worked off a crank shaft. At each downward plunge, the water is forced upwards through the perforated copper grating, causing the breeze and clinkers to rise, when, owing to the greater specific gravity of the clinker, on the return stroke it precipitates to the bottom; at the same time, the breeze or unburnt fuel, being lighter, works to the surface; and at each stroke of the crank a body of water and quantity of clean breeze is carried on to a perforated plate, allowing the water to return to the machine to be used over again. The clean breeze is swept off this plate by a revolving brush. The clinker accumulates on the copper plate, and ultimately falls into the body of the machine, whence it is thrown out by an elevator.

In our illustration (Fig. 59), C is the outlet for clinker, E is the elevator, R the revolving brush, and B the outlet for breeze. These machines can be driven either by hand or power, according to size; and the inventor states that cases are on record where firms on shore have been paying two shillings a load to cart away ashes that produced by means of this machine as much as 45 per cent. of their weight in useful fuel, a result that must have been very gratifying to both purchaser and inventor.

Ashpits should be always kept clear of ashes, otherwise there will be a deficient supply of air and a consequent loss of draught; besides, if they are allowed to accumulate to any great extent the fire-bars may become red-hot, and probably come down. Should there be any difficulty experienced in keeping steam from any cause whatever, it is more economical to link up the expansion or valve gear, until the full pressure can be maintained, than to have everything full open with a lower pressure.

In vessels not fitted with evaporators or other appliances for condensing water to supply the waste in the boilers, incrustation is bound to take place, consequently every

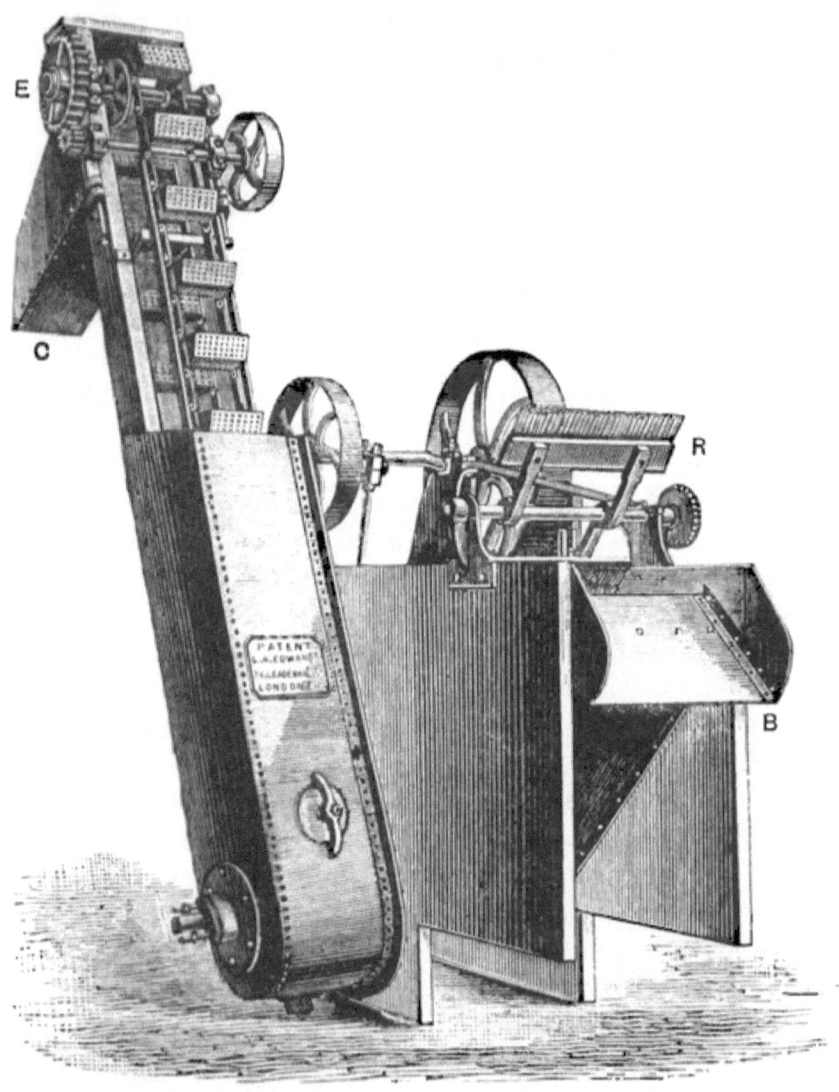

Fig. 59.—Ashes Washing Machine.

precaution ought to be taken to reduce this evil as much as possible. No steam must be wasted, and the glands and safety valves should be kept perfectly tight.

Fires should always be banked at the front and not at the back, as it has been maintained that furnace fronts have frequently become leaky through the latter practice, the cold air acting on the fronts causing unequal expansion. Keep the ashpit dampers shut, and see that the smoke-box door and uptake are air-tight. The smoke-box doors should always be kept closed when under steam, and care taken to shut the furnace doors immediately after firing.

After arriving in port, and when finished with steam, allow the fires to die out, keeping the furnace doors and dampers closed. Do not blow the boiler down, for if it is blown down when hot great injury is likely to ensue from too sudden contraction, but when the water is comparatively cold it may be allowed to run into the bilges. Let the boiler dry thoroughly after the water is run out, and do not fill it again until just before the fires require to be lighted.

The boilers, steam pipes, and cylinders should be kept well covered to prevent loss of heat by radiation, and any part becoming exposed should be re-covered without delay.

There are numerous other minor matters in the management of triple and quadruple expansion engines and boilers that require attention, but as they are equally necessary with other engines, and must have already come under the notice of every intelligent and careful engineer who has had experience with ordinary marine engines, they need not be gone into here ; but it may be added, by way of encouragement, that the more care and attention bestowed by an engineer upon the engines and boilers committed to his care, the more amply will he be repaid through the freedom from worry and annoyance that can only be realised by keeping machinery in the highest condition of efficiency, and by the consciousness of the fact that he has faithfully discharged his duty to those who have entrusted him with much valuable property.

CHAPTER X.

Specification of Triple Expansion Marine Engines—Cylinders—Cylinder Covers—Pistons—Slide Valves—Valve Gear—Piston-Rods—Connecting-Rods—Columns—Air-Pump—Circulating-Pump—Feed-Pumps—Bilge-Pumps — Condenser — Evaporator—Drain Trap—Bed-plate—Crankshaft—Shafting—Propeller—Turning Gear—Telegraph—Starting Gear—Throttle Valve—Donkey Engine—Sanitary Pump—Ballast Engine — Pipes, Lubricators, etc.—Conditions—Boilers—Tubes—Staying—Funnel and Uptake—Mountings—Ventilators.

ALTHOUGH an engineer is expected to be quite conversant with every detail of modern marine engines and boilers, it is very seldom indeed that he has an opportunity of seeing the actual specification, containing the dimensions and other conditions under which the machinery under his care has been constructed; and it may consequently help to make his knowledge of such matters more complete if he has placed before him a typical specification of the present day.

The following copy of a specification for triple expansion engines and boilers for a steamer of ——— tons ——— indicated horse-power, and similar in most respects to the P. & O. Company's s.s. *Aden*, has been kindly furnished by Messrs. Richardson of Hartlepool, the builders of that vessel, for insertion in this work.

SPECIFICATION OF TRIPLE EXPANSION ENGINES.

Cylinders.

The cylinders to be three in number:—The first or high-pressure cylinder to be 28 inches diameter. The second or intermediate cylinder to be 46 inches diameter. The third or low-pressure cylinder to be 77 inches diameter. The stroke to be 4 feet in each case. All the cylinders to be thoroughly sound castings of hard close-grained iron, and fitted with working liners smoothly and truly turned and bored. The high-pressure cylinder to have a piston slide valve arranged to take steam at the centre, and exhaust at the ends. The piston valve to have loose working liners secured by flanges and studs. The intermediate and low-pressure cylinders to have flat slide valves, and loose working faces secured by Muntz metal pins, recessed so that the heads are $\frac{1}{8}$ inch below the flush. The face to be five-ported.

The high-pressure and intermediate cylinders to be lagged with non-conducting cement 2 inches thick, and covered with sheet-iron. Low-pressure cylinder to be covered with felt and sheet-iron. The cylinders to be fitted with all necessary drain cocks, and pipes led to condenser, bilge, and drain tank. All drains to be worked from the starting platform. Relief valves to be fitted to top and bottom of cylinders, also indicator cocks and suitable gear.

Cylinder Covers.

To be of double form, having strong ribs, and to be chequered on top, to have lugs and shackles on cylinder, and valve covers with suitable lifting gear, and strong beam fitted with suitable travellers. Grease cocks to be fitted to IP and LP cylinder covers.

Pistons.

The pistons to be of double form, having metallic packing. The HP and IP to be fitted with Ramsbottom's rings, and the LP with a packing-ring and coil springs.

The junk-ring studs to be of wrought-iron $1\frac{1}{4}$ inch diameter with the nuts secured by stout steel pins. A square to be formed on studs close to piston, and the holes on junk-rings recessed to suit.

Slide Valves.

The HP cylinder piston valve to be fitted with approved packing-rings. The rings to form the cutting-off edges, both for steam and exhaust. The IP and LP slide valves to be double ported. All valves to be accurately fitted.

Valve spindles to be guided at top by brass bushes, and at bottom by strong adjustable guide bracket of large surface.

Valve Gear.

To have long bar-links of steel. Working parts to have phosphor-bronze bearings of large surface, all to be adjustable. Weigh-shaft to have caps, so that wear can be taken up. Eccentric-pulleys of cast-iron with cast-brass liners. Eccentric-rods and straps of wrought iron with adjustable ends and large bearing surface. An adjustable expansion arrangement to be fitted to each cylinder.

PISTON-RODS.

The piston-rods to be of mild ingot steel, all to be $8\frac{1}{2}$ inches in diameter, to be guided below by adjustable cast-iron shoes fitted on ahead side with white metal of large surface, to have gudgeons fitted to piston-rods for working the pumps.

Metallic packing to be fitted to HP piston-rod gland.

All piston-rod glands to be fitted with geared nuts and wheel for uniform tightening, and to have locking arrangement for securing the gland.

CONNECTING-RODS.

The connecting-rods to be of the best hammered scrap iron, 9 feet 6 inches between centres, fitted with adjustable ends—the top phosphor-bronze and the bottom of strong cast-iron with white metal strips.

COLUMNS.

The front columns to be of cast-iron of the inverted Y box form, and arranged as oil tanks having lock-cocks and sludging-doors fitted.

The guide plates to be bolted to the columns. The go-ahead guide to be fitted for water to circulate at the back.

Feet of columns to be of large surface securely bolted to bed-plates and cylinders by turned and fitted bolts.

AIR-PUMPS.

The air-pump to be worked by levers connected with the after engine, and of sufficient capacity for working by common injection. The working barrel, bucket, valve seats and guards to be of brass, and the pump-rod to be of Muntz metal. The air-pump rod to be guided by means of a bracket with adjustable bushes fitted to the condenser.

Thompson's metallic disc valves to be supplied and so arranged as to be capable of being replaced without removing the crosshead.

CIRCULATING-PUMP.

The circulating-pump to be double acting, and worked by the same levers as the air-pump; the working barrel, bucket, valve seats, and guards to be of brass; the pump-rod to be of Muntz metal and the valves of india-rubber or vulcanised fibre.

A bilge injection pipe to be fitted to the circulating-pump.

FEED-PUMPS.

To have two feed-pumps of cast-iron with plungers and valves of brass, each capable of supplying the boiler with sufficient water when the engines are working at full power. Each pump to be fitted with a gun-metal escape valve with adjustable spiral spring, and with separate cocks or valves on the suction pipes. Each pump also to have a non-return self-acting valve on the discharge branch to enable the valves of one pump to be examined while the other pump is working. Air vessels to be fitted to the pumps.

BILGE-PUMPS.

To have two bilge-pumps of cast-iron with valves and valve seats of brass, and having stop-cocks or valves fitted on suction and self-acting check-valves on delivery pipes between the pumps and ship's side. A bilge suction chest to be fitted in a convenient position in the engine-room, and fitted with valve connections for each compartment of the ship.

One bilge-pump to be arranged to draw from the sea, and discharge on deck.

Condenser.

The condenser to be of cast-iron, and arranged to form part of the framing for supporting the cylinders, to be firmly secured to the bed-plate and cylinders by means of turned and fitted bolts, the air and circulating pumps to be bolted to the condenser. Tubes to be of solid drawn brass, secured in brass tube plates 1 inch thick, by brass screwed ferrules with cotton packing. The cooling surface not to be less than 4200 square feet, to have man and sludge doors, soda-cocks and auxiliary feed-cock,—all bolts exposed to the action of the water to be of Muntz metal with brass nuts.

Evaporator.

Morison's patent evaporator and feed-heater of ample capacity to be supplied and fitted. The feed-heater to be placed on the feed suction pipe, so that all the water from the hot well to the feed-pumps passes through the feed-heater, and is heated by the steam generated in the co-operator.

Drain Trap.

Two pulsator automatic drain traps (Morison's patent) to be fitted to the IP and LP receivers, to automatically remove any water that may accumulate.

BED-PLATE.

The bed-plate to be a strong casting on the box system, in three pieces, securely bolted together by turned and fitted bolts, and having strong pedestals for receiving the shaft-bearings.

Main-bearing brushes to be of strong cast-iron, lined with white metal, and fitted with polished wrought-iron caps secured by iron bolts and nuts.

The bed-plate to have extra strong flanges for bolting to the engine seating.

CRANKSHAFT.

The crankshaft to be a built shaft made in three pieces, with flanged couplings on each end; each piece to be an exact duplicate of the other.

The shafts and pins to be of ingot steel, without mark or flaw, and the webs to be of best selected scrap-iron. The three shafts to be bolted together, and to receive a finishing cut in the lathe.

SHAFTING.

The propeller-shaft to be fitted with heavy solid brass liners. The tunnel bearings to be of cast-iron, with shell covers, and lined with white metal on the bottom.

A horse-shoe thrust-block of large surface (lined with white metal) to be fitted with a plain bearing immediately behind it, but separate and adjustable.

An efficient water service to be provided for all bearings.

The bearing at outer end of stern-tube to be of brass,

with lignum vitæ strips not less than 56 inches long. Inner bearing of brass.

Stern-tube neck bush of brass, and gland bushed with brass, with iron studs and brass nuts.

Propeller.

The propeller to have adjustable blades. Box to be of cast-iron, of spherical form, with four cast-steel blades with oval holes to vary pitch 3 feet; studs of steel with brass-capped nuts and lock pins; propeller, 18 feet diameter; pitch variable, from 22 to 25 feet; surface not less than 98 square feet.

The propeller-blades to be sheathed with brass on back over one-third of their area from the tip.

Turning Gear.

The turning gear to consist of a powerful worm and wheel fitted on shaft in engine-room near to the bulkhead, and to be worked both by hand and from the ballast-donkey, and to be arranged so that it can be conveniently disengaged.

Telegraph.

To have a Chadburn's repeating telegraph from bridge and flying bridge to engine-room, with lamps, etc. A speaking-tube to be fitted from starting platform to the captain's bridge.

Starting Gear.

To have a powerful starting gear, to be worked by steam and hand. Auxiliary starting valves to be fitted to the first and second receivers.

Throttle Valve, etc.

To have a butterfly throttle valve; also a strong stop-valve worked from the starting platform, and placed immediately behind the throttle valve.

Donkey-Engine.

To have a double-acting donkey-engine, 9 inch cylinder, $4\frac{1}{2}$ inch pump, to take steam from main and donkey boilers, and to exhaust to winch tank and condenser. The donkey to draw from sea, bilge, and hot well, and to deliver on deck, overboard, through condenser, and into boilers. To have separate feed-pipes and check-valves from main feed-pumps, and to supply donkey boiler and sanitary tanks.

Sanitary Pump.

A sanitary pump to be fitted, worked by main engines, to draw from the sea, and to deliver into tanks on deck. The pump to be strongly constructed; all pins of large surface and all bearings adjustable.

Ballast Engine.

To have a double-acting water ballast engine, 10 inch cylinder, $5\frac{1}{2}$ inch pump, to pump from tank or bilges, and to deliver overboard or through condenser.

Pipes, Lubricators, etc.

To have a complete set of pipes for a working pressure of 160 lbs. Main steam-pipe not less than $\frac{5}{16}$ inch thick. Bilge pipes of lead. Bilge discharge pipes of strong cast-iron.

All necessary relief valves, lubricators, grease cups, etc. to be provided; also one Mollerup's lubricator, connected to HP valve-casing.

Ladders, platforms, and handrails in engine-room and stokehold as may be required for the convenient working of the engines.

Stokehold plates of chequered cast-iron, with strong wood platform under them.

The engines to have three coats of paint of approved colour, and one of varnish.

The engine-room to be neatly painted, grained, and varnished.

CONDITIONS.

The whole of the machinery to be of the best materials and workmanship, carefully fitted on board ship, and finished in a first-class manner, in accordance with this specification.

The engines to be afterwards tried at moorings and at sea, and proved satisfactorily efficient in all respects.

The engines and boilers to be guaranteed against all breakages and defects arising from imperfect workmanship or bad materials for six months after completion.

SPECIFICATION FOR TWO DOUBLE-ENDED BOILERS.

To be of mild steel, 15 feet 3 inches diameter, 17 feet long, to pass Board of Trade and Lloyds' Survey for a working pressure of 160 lbs. To have in all 12 Morison's suspension furnaces, made of mild steel of a tensile strength, not exceeding 26 tons per square inch. Diameter of furnaces, 45 inches.

All butts and seams to be kept from the action of the fire where practicable.

Drawings of boilers to be submitted for approval before constructing.

Richardson's improved doors to each furnace.

Boilers to be covered with non-conducting cement, neatly secured with galvanised wrought-iron sheets.

Zinc plates to be fitted in boilers, to prevent galvanic action.

TUBES AND STAY TUBES.

Stay tubes to be screwed into back tube-plates, and to have nuts outside and inside front tube-plates.

Plain tubes $3\frac{1}{4}$ inches diameter, $4\frac{1}{2}$ inches pitch, to project $\frac{5}{8}$ inches from front tube-plates. Back ends of tube to be leaded.

STAYING.

The staying to be arranged to meet the requirements of the Board of Trade and Lloyds, for a working pressure of 160 lbs.

FUNNEL AND UPTAKES.

The funnel to be of ample size (about 20 per cent. in excess of the tube area). The front of uptakes to be fitted with baffle plates, in accordance with drawings submitted.

Damper to be fitted in funnel uptake, worked from engine-room.

All necessary manhole and manhole-doors, furnace-doors, fire-bars, bearers, dead-plates, etc., to be supplied and fitted in a first-class manner.

MOUNTINGS.

Each boiler to have one stop-valve and two spring safety valves, with suitable easing gear, worked from engine-room.

All feed-check valves, water-gauge, and test-cocks to have brass-packed plugs.

Internal feed-pipes to deliver between tubes.

Blow-off cocks to be fitted in line of centre wing furnace.

Test and salinometer cocks to be fitted to each boiler.

VENTILATORS.

Two ventilators to be placed in each stokehold, with shifting cowls and gear worked from the stokehold. An ash hoist to be fitted in each stokehold.

www.ingramcontent.com/pod-product-compliance
Lightning Source LLC
Chambersburg PA
CBHW021346230426
43666CB00006B/429